# CAMBRIDGE LIBRARY COLLECTION

*Books of enduring scholarly value*

## Physical Sciences

From ancient times, humans have tried to understand the workings of the world around them. The roots of modern physical science go back to the very earliest mechanical devices such as levers and rollers, the mixing of paints and dyes, and the importance of the heavenly bodies in early religious observance and navigation. The physical sciences as we know them today began to emerge as independent academic subjects during the early modern period, in the work of Newton and other 'natural philosophers', and numerous sub-disciplines developed during the centuries that followed. This part of the Cambridge Library Collection is devoted to landmark publications in this area which will be of interest to historians of science concerned with individual scientists, particular discoveries, and advances in scientific method, or with the establishment and development of scientific institutions around the world.

## Conversations on Natural Philosophy

An author of educational works intended especially for young women, Jane Haldimand Marcet (1769–1858) sought to combat the notion that technical topics were unsuitable for female students. Inspired by conversations with the famous scientists she entertained, she wrote textbooks in the lively form of discussions between a teacher and her two female pupils. Published anonymously at first, they found broad popularity: Michael Faraday, as a young bookbinder's apprentice, credited Marcet with introducing him to electrochemistry. The present work, an introduction to physics, astronomy and the properties of matter, sound and light, was Marcet's first, though it remained unpublished until 1819. Her other works include *Conversations on Chemistry* (1805), *Conversations on Political Economy* (1816) and *Conversations on Vegetable Physiology* (1829), all of which are reissued in this series. Never professing to be original, Marcet's work is noted nonetheless for its thoroughness and clear presentation of concepts.

Cambridge University Press has long been a pioneer in the reissuing of out-of-print titles from its own backlist, producing digital reprints of books that are still sought after by scholars and students but could not be reprinted economically using traditional technology. The Cambridge Library Collection extends this activity to a wider range of books which are still of importance to researchers and professionals, either for the source material they contain, or as landmarks in the history of their academic discipline.

Drawing from the world-renowned collections in the Cambridge University Library and other partner libraries, and guided by the advice of experts in each subject area, Cambridge University Press is using state-of-the-art scanning machines in its own Printing House to capture the content of each book selected for inclusion. The files are processed to give a consistently clear, crisp image, and the books finished to the high quality standard for which the Press is recognised around the world. The latest print-on-demand technology ensures that the books will remain available indefinitely, and that orders for single or multiple copies can quickly be supplied.

The Cambridge Library Collection brings back to life books of enduring scholarly value (including out-of-copyright works originally issued by other publishers) across a wide range of disciplines in the humanities and social sciences and in science and technology.

# Conversations on Natural Philosophy

*In which the Elements of that Science are Familiarly Explained and Adapted to the Comprehension of Young Pupils*

JANE HALDIMAND MARCET

CAMBRIDGE
UNIVERSITY PRESS

# CAMBRIDGE
## UNIVERSITY PRESS

University Printing House, Cambridge, CB2 8BS, United Kingdom

Published in the United States of America by Cambridge University Press, New York

Cambridge University Press is part of the University of Cambridge.
It furthers the University's mission by disseminating knowledge in the pursuit of
education, learning and research at the highest international levels of excellence.

www.cambridge.org
Information on this title: www.cambridge.org/9781108067010

© in this compilation Cambridge University Press 2014

This edition first published 1819
This digitally printed version 2014

ISBN 978-1-108-06701-0 Paperback

# CONVERSATIONS

## ON

# NATURAL PHILOSOPHY,

IN WHICH

## THE ELEMENTS OF THAT SCIENCE

ARE

## *FAMILIARLY EXPLAINED,*

AND ADAPTED TO THE

## COMPREHENSION OF YOUNG PUPILS.

ILLUSTRATED WITH PLATES.

BY THE AUTHOR OF

*CONVERSATIONS ON CHEMISTRY,*

AND

*CONVERSATIONS ON POLITICAL ECONOMY.*

*LONDON:*

PRINTED FOR LONGMAN, HURST, REES, ORME, AND BROWN,

PATERNOSTER-ROW.

1819.

Printed by Strahan and Spottiswoode,
Printers-Street, London.

# PREFACE.

It is with increased diffidence that the author offers this little work to the public. The encouraging reception which the Conversations on Chemistry, and Political Economy have met with, has induced her to venture on publishing a short course on Natural Philosophy; but not without the greatest apprehensions for its success. Her ignorance of mathematics, and the imperfect knowledge of natural philosophy which that disadvantage necessarily implies, renders her fully sensible of her incompetency to treat the subject in any other way than in the form of a familiar explanation of the first elements, for the use of very young pupils. It is the hope of having done this in a manner that may engage their attention, which encourages her to offer them these additional lessons.

They are intended, in a course of elementary science, to precede the Conversations on Chemistry; and were actually written previous to either of her former publications.

# CONTENTS.

A 3

## CONVERSATION XVII.

# DIRECTIONS FOR PLACING THE ENGRAVINGS.

---

ERRATUM.

Page 98. for Plate III. read Plate IV.

# CONVERSATION I.

## ON GENERAL PROPERTIES OF BODIES.

INTRODUCTION.—GENERAL PROPERTIES OF BODIES.
—IMPENETRABILITY.— EXTENSION.—FIGURE.—
DIVISIBILITY. — INERTIA. — ATTRACTION. — AT-
TRACTION OF COHESION.— DENSITY.— RARITY.
— HEAT. — ATTRACTION OF GRAVITATION.

### EMILY.

I MUST request your assistance, my dear Mrs. B.,
in a charge which I have lately undertaken: it is
that of instructing my youngest sister, a task, which I
find proves more difficult than I had at first imagined.
I can teach her the common routine of children's
lessons tolerably well; but she is such an inquisi-
tive little creature, that she is not satisfied without
an explanation of every difficulty that occurs to her,
and frequently asks me questions which I am at a
loss to answer. This morning, for instance, when
I had explained to her that the world was round

B

like a ball, instead of being flat as she had supposed, and that it was surrounded by the air, she asked me what supported it. I told her that it required no support; she then enquired why it did not fall as every thing else did? This I confess perplexed me; for I had myself been satisfied with learning that the world floated in the air, without considering how unnatural it was that so heavy a body, bearing the weight of all other things, should be able to support itself.

## MRS. B.

I make no doubt, my dear, but that I shall be able to explain this difficulty to you; but I believe that it would be almost impossible to render it intelligible to the comprehension of so young a child as your sister Sophia. You, who are now in your thirteenth year, may, I think, with great propriety, learn not only the cause of this particular fact, but acquire a general knowledge of the laws by which the natural world is governed.

## EMILY.

Of all things, it is what I should most like to learn; but I was afraid it was too difficult a study even at my age.

## MRS. B.

Not when familiarly explained: if you have patience to attend, I will most willingly give you all the information in my power. You may per-

haps find the subject rather dry at first; but if I
succeed in explaining the laws of nature, so as to
make you understand them, I am sure that you
will derive not only instruction, but great amusement
from that study.

EMILY.

I make no doubt of it, Mrs. B.; and pray begin
by explaining why the earth requires no support;
for that is the point which just now most strongly
excites my curiosity.

MRS. B.

My dear Emily, if I am to attempt to give you a
general idea of the laws of nature, which is no less
than to introduce you to a knowledge of the science
of natural philosophy, it will be necessary for us to
proceed with some degree of regularity. I do not
wish to confine you to the systematic order of a
scientific treatise; but if we were merely to examine
every vague question that may chance to occur, our
progress would be but very slow. Let us, therefore,
begin by taking a short survey of the general properties of bodies, some of which must necessarily
be explained before I can attempt to make you understand why the earth requires no support.

When I speak of *bodies*, I mean substances, of
whatever nature, whether solid or fluid; and *matter*
is the general term used to denote the substance,
whatever its nature be, of which the different bodies

are composed. Thus, wood is the matter of which this table is made; water is the matter with which this glass is filled, &c.

### EMILY.

I am very glad you have explained the meaning of the word matter, as it has corrected an erroneous conception I had formed of it: I thought that it was applicable to solid bodies only.

### MRS. B.

There are certain properties which appear to be common to all bodies, and are hence called the *essential properties* of bodies; these are, *Impenetrability, Extension, Figure, Divisibility, Inertia,* and *Attraction*. These are called the general properties of bodies, as we do not suppose any body to exist without them.

By impenetrability, is meant the property which bodies have of occupying a certain space, so that, where one body is, another cannot be, without displacing the former; for two bodies cannot exist in the same place at the same time. A liquid may be more easily removed than a solid body; yet it is not the less substantial, since it is as impossible for a liquid and a solid to occupy the same space at the same time, as for two solid bodies to do so. For instance, if you put a spoon into a glass full of water, the water will flow over to make room for the spoon.

EMILY.

I understand this perfectly. Liquids are in reality as substantial or as impenetrable as solid bodies, and they appear less so, only because they are more easily displaced.

MRS. B.

The air is a fluid differing in its nature from liquids, but no less impenetrable. If I endeavour to fill this phial by plunging it into this bason of water, the air, you see, rushes out of the phial in bubbles, in order to make way for the water, for the air and the wate cannot exist together in the same space, any more than two hard bodies; and if I reverse this goblet, and plunge it perpendicularly into the water, so that the air will not be able to escape, the water will no longer be able to fill the goblet.

EMILY.

But it rises a considerable way into the glass.

MRS. B.

Because the water compresses or squeezes the air into a small space in the upper part of the glass; but, as long as it remains there, no other body can occupy the same place.

EMILY.

A difficulty has just occurred to me, with regard to the impenetrability of solid bodies; if a nail is

B 3

driven into a piece of wood, it penetrates it, and both the wood and the nail occupy the same space that the wood alone did before?

MRS. B.

The nail penetrates between the particles of the wood, by forcing them to make way for it; for you know that not a single atom of wood can remain in the space which the nail occupies; and if the wood is not increased in size by the addition of the nail, it is because wood is a porous substance, like sponge, the particles of which may be compressed or squeezed closer together; and it is thus that they make way for the nail.

We may now proceed to the next general property of bodies, *extension.* A body which occupies a certain space must necessarily have extension; that is to say, *length, breadth,* and *depth ;* these are called the dimensions of extension: can you form an idea of any body without them?

EMILY.

No; certainly I cannot; though these dimensions must, of course, vary extremely in different bodies. The length, breadth, and depth of a box, or of a thimble, are very different from those of a walking-stick, or of a hair.

But is not height also a dimension of extension?

MRS. B.

Height and depth are the same dimension, considered in different points of view; if you measure a body, or a space, from the top to the bottom, you call it depth; if from the bottom upwards, you call it height; thus the depth and height of a box are, in fact, the same thing.

EMILY.

Very true; a moments consideration would have enabled me to discover that; and breadth and width are also the same dimension.

MRS. B.

Yes; the limits of extension constitute *figure* or shape. You conceive that a body having length, breadth, and depth, cannot be without form, either symmetrical or irregular?

EMILY.

Undoubtedly; and this property admits of almost an infinite variety.

MRS. B.

Nature has assigned regular forms to her productions in general. The natural form of mineral substances is that of crystals, of which there is a great variety. Many of them are very beautiful, and no less remarkable by their transparency, or colour, than by the perfect regularity of their forms,

as may be seen in the various museums and collections of natural history. The vegetable and
animal creation appears less symmetrical, but is
still more diversified in figure than the mineral
kingdom. Manufactured substances assume the
various arbitrary forms which the art of man designs
for them; and an infinite number of irregular forms
are produced by fractures, and by the dismember-
ment of the parts of bodies.

EMILY.

Such as a piece of broken china, or glass?

MRS. B.

Or the fragments of mineral bodies which are
broken in being dug out of the earth, or decayed
by the effect of torrents and other causes. The
picturesque effect of rock-scenery is in a great
measure owing to accidental irregularities of this
kind.

We may now proceed to divisibility; that is to
say, a susceptibility of being divided into an inde-
finite number of parts. Take any small quantity
of matter, a grain of sand for instance, and cut it
into two parts; these two parts might be again
divided, had we instruments sufficiently fine for the
purpose; and if, by means of pounding, grinding,
and other similar methods, we carry this division to
the greatest possible extent, and reduce the body

to its finest imaginable particles, yet not one of the particles will be destroyed, and the body will continue to exist, though in this altered state.

The melting of a solid body in a liquid affords a very striking example of the extreme divisibility of matter; when you sweeten a cup of tea, for instance, with what minuteness the sugar must be divided to be diffused throughout the whole of the liquid.

EMILY.

And if you pour a few drops of red wine into a glass of water, they immediately tinge the whole of the water, and must therefore be diffused throughout it.

MRS. B.

Exactly so; and the perfume of this lavender-water will be almost as instantaneously diffused throughout the room, if I take out the stopper.

EMILY.

But in this case it is only the perfume of the lavender, and not the water itself, that is diffused in the room?

MRS. B.

The odour or smell of a body is part of the body itself, and is produced by very minute particles or exhalations which escape from odoriferous bodies. It would be impossible that you should smell the lavender-water, if particles of it did not come in actual contact with your nose.

EMILY.

But when I smell a flower, I see no vapour rise from it; and yet I can perceive the smell at a considerable distance.

MRS. B.

You could, I assure you no more smell a flower, the odoriferous particles of which did not touch your nose, than you could taste a fruit, the flavoured particles of which did not come in contact with your tongue.

EMILY.

That is wonderful indeed; the particles then, which exhale from the flower and from the lavender-water, are, I suppose, too small to be visible?

MRS. B.

Certainly: you may form some idea of their extreme minuteness, from the immense number which must have escaped in order to perfume the whole room; and yet there is no sensible diminution of the liquid in the phial.

EMILY.

But the quantity must really be diminished?

MRS. B.

Undoubtedly; and were you to leave the bottle

open a sufficient length of time, the whole of the water would evaporate and disappear. But though so minutely subdivided as to be imperceptible to any of our senses, each particle would continue to exist; for it is not within the power of man to destroy a single particle of matter; nor is there any reason to suppose that in nature an atom is ever annihilated.

### EMILY.

Yet, when a body is burnt to ashes, part of it, at least, appears to be effectually destroyed? Look how small is the residue of ashes beneath the grate, from all the coals which have been consumed within it.

### MRS. B.

That part of the coals, which you suppose to be destroyed, evaporates in the form of smoke and vapour, whilst the remainder is reduced to ashes. A body, in burning, undergoes no doubt very remarkable changes; it is generally subdivided; its form and colour altered; its extension increased: but the various parts, into which it has been separated by combustion, continue in existence, and retain all the essential properties of bodies.

### EMILY.

But that part of a burnt body which evaporates in smoke has no figure; smoke, it is true, ascends in columns into the air, but it is soon so much diffused as to lose all form; it becomes indeed invisible.

MRS. B.

Invisible, I allow; but we must not imagine that what we no longer see no longer exists. Were every particle of matter that becomes invisible annihilated, the world itself would in the course of time be destroyed. The particles of smoke, when diffused in the air, continue still to be particles of matter, as well as when more closely united in the form of coals: they are really as substantial in the one state as in the other, and equally so when by their extreme subdivision they become invisible. No particle of matter is ever destroyed: this is a principle you must constantly remember. Every thing in nature decays and corrupts in the lapse of time. We die, and our bodies moulder to dust; but not a single atom of them is lost; they serve to nourish the earth, whence, while living, they drew their support.

The next essential property of matter is called *inertia;* this word expresses the resistance which inactive matter makes to a change of state. Bodies appear to be equally incapable of changing their actual state, whether it be of motion or of rest. You know that it requires force to put a body which is at rest in motion; an exertion of strength is also requisite to stop a body which is already in motion.

The resistance of the body to a change of state, in either case, is called its *inertia*.

EMILY.

In playing at base-ball I am obliged to use all my strength to give a rapid motion to the ball; and when I have to catch it, I am sure I feel the resistance it makes to being stopped.   But if I did not catch it, it would soon fall to the ground and stop of itself.

MRS. B.

Inert matter is as incapable of stopping of itself, as it is of putting itself into motion: when the ball ceases to move, therefore, it must be stopped by some other cause or power; but as it is one with which you are yet unacquainted, we cannot at present investigate its effects.

The last property which appears to be common to all bodies is *attraction*.   All bodies consist of infinitely small particles of matter, each of which possesses the power of attracting or drawing towards it, and uniting with any other particle sufficiently near to be within the influence of its attraction; but in minute particles this power extends to so very small a distance around them, that its effect is not sensible, unless they are (or at least appear to be) in contact; it then makes them stick or adhere together, and is hence called the *attraction of cohesion*.   Without this power, solid bodies would fall in pieces, or rather crumble to atoms.

EMILY.

I am so much accustomed to see bodies firm and solid, that it never occurred to me that any power was requisite to unite the particles of which they are composed. But the attraction of cohesion does not, I suppose, exist in liquids; for the particles of liquids do not remain together so as to form a body, unless confined in a vessel?

MRS. B.

I beg your pardon; it is the attraction of cohesion which holds this drop of water suspended at the end of my finger, and keeps the minute watery particles of which it is composed united. But as this power is stronger in proportion as the particles of bodies are more closely united, the cohesive attraction of solid bodies is much greater than that of fluids.

The thinner and lighter a fluid is, the less is the cohesive attraction of its particles, because they are further apart; and in elastic fluids, such as air, there is no cohesive attraction among the particles.

EMILY.

That is very fortunate; for it would be impossible to breathe the air in a solid mass; or even in a liquid state.

But is the air a body of the same nature as other bodies?

#### MRS. B.

Undoubtedly, in all essential properties.

#### EMILY.

Yet you say that it does not possess one of the general properties of bodies — cohesive attraction?

#### MRS. B.

The particles of air are not destitute of the power of attraction, but they are too far distant from each other to be influenced by it; and the utmost efforts of human art have proved ineffectual in the attempt to compress them, so as to bring them within the sphere of each other's attraction, and make them cohere.

#### EMILY.

If so, how is it possible to prove that they are endowed with this power?

#### MRS. B.

The air is formed of particles precisely of the same nature as those which enter into the composition of liquid and solid bodies, in which state we have a proof of their attraction.

#### EMILY.

It is then, I suppose, owing to the different degrees of attraction of different substances, that they are hard or soft; and that liquids are thick or thin?

MRS. B.

Yes; but you would express your meaning better by the term *density*, which denotes the degree of closeness and compactness of the particles of a body: thus you may say, both of solids, and of liquids, that the stronger the cohesive attraction, the greater is the density of the body.   In philosophical language, density is said to be that property of bodies by which they contain a certain quantity of matter, under a certain bulk or magnitude.   *Rarity* is the contrary of density; it denotes the thinness and subtlety of bodies: thus you would say that mercury or quicksilver was a very dense fluid; ether, a very rare one, &c.

CAROLINE.

But how are we to judge of the quantity of matter contained in a certain bulk?

MRS. B.

By the weight: under the same bulk bodies are said to be dense in proportion as they are heavy.

EMILY.

Then we may say that metals are dense bodies, wood comparatively a rare one, &c.   But, Mrs. B., when the particles of a body are so near as to at-

attract each other, the effect of this power must increase as they are brought by it closer together; so that one would suppose that the body would gradually augment in density, till it was impossible for its particles to be more closely united. Now, we know that this is not the case; for soft bodies, such as cork, sponge, or butter, never become, in consequence of the increasing attraction of their particles, as hard as iron?

MRS. B.

In such bodies as cork and sponge, the particles which come in contact are so few as to produce but a slight degree of cohesion: they are porous bodies, which, owing to the peculiar arrangement of their particles, abound with interstices which separate the particles; and these vacancies are filled with air, the spring or elasticity of which prevents the closer union of the parts. But there is another fluid much more subtle than air, which pervades all bodies, this is *heat*. Heat insinuates itself more or less between the particles of all bodies, and forces them asunder; you may therefore consider heat, and the attraction of cohesion, as constantly acting in opposition to each other.

EMILY.

The one endeavouring to rend a body to pieces, the other to keep its parts firmly united.

MRS. B.

And it is this struggle between the contending forces of heat and attration, which prevents the extreme degree of density which would result from the sole influence of the attraction of cohesion.

EMILY.

The more a body is heated then, the more its particles will be separated.

MRS. B.

Certainly: we find that bodies swell or dilate by heat: this effect is very sensible in butter, for instance, which expands by the application of heat, till at length the attraction of cohesion is so far diminished that the particles separate, and the butter becomes liquid. A similar effect is produced by heat on metals, and all bodies susceptible of being melted. Liquids, you know, are made to boil by the application of heat; the attraction of cohesion then yields entirely to the expansive power; the particles are totally separated and converted into steam or vapour. But the agency of heat is in no body more sensible than in air, which dilates and contracts by its increase or diminution in a very remarkable degree.

EMILY.

The effects of heat appear to be one of the most interesting parts of natural philosophy.

That is true; but heat is so intimately connected with chemistry, that you must allow me to defer the investigation of its properties till you become acquainted with that science.

To return to its antagonist, the attraction of cohesion; it is this power which restores to vapour its liquid form, which unites it into drops when it falls to the earth in a shower of rain, which gathers the dew into brilliant gems on the blades of grass.

EMILY.

And I have often observed that after a shower, the water collects into large drops on the leaves of plants; but I cannot say that I perfectly understand how the attraction of cohesion produces this effect.

MRS. B.

Rain does not fall from the clouds in the form of drops, but in that of mist or vapour, which is composed of very small watery particles; these, in their descent, mutually attract each other, and those that are sufficiently near in consequence unite and form a drop, and thus the mist is transformed into a shower. The dew also was originally in a state of vapour, but is, by the mutual attraction of the particles, formed into small globules on the blades of grass: in a similar manner the rain upon the leaf collects into large drops, which when they become too heavy for the leaf to support fall to the ground.

EMILY.

All this is wonderfully curious ! I am almost bewildered with surprise and admiration at the number of new ideas I have already acquired.

MRS. B.

Every step that you advance in the pursuit of natural science, will fill your mind with admiration and gratitude towards its Divine Author. In the study of natural philosophy, we must consider ourselves as reading the book of nature, in which the bountiful goodness and wisdom of God is revealed to all mankind; no study can then tend more to purify the heart, and raise it to a religious contemplation of the Divine perfections.

There is another curious effect of the attraction of cohesion which I must point out to you. It enables liquids to rise above their level in capillary tubes: these are tubes the bores of which are so extremely small that liquids ascend within them, from the cohesive attraction between the particles of the liquid and the interior surface of the tube. Do you perceive the water rising above its level in this small glass tube, which I have immersed in a goblet full of water?

EMILY.

Oh yes; I see it slowly creeping up the tube, but now it is stationary: will it rise no higher?

MRS. B.

No; because the cohesive attraction between the water and the internal surface of the tube is now balanced by the weight of the water within it: if the bore of the tube were narrower the water would rise higher; and if you immerse several tubes of bores of different sizes, you will see it rise to different heights in each of them. In making this experiment you should colour the water with a little red wine, in order to render the effect more obvious.

All porous substances, such as sponge, bread, linen, &c., may be considered as collections of capillary tubes: if you dip one end of a lump of sugar into water, the water will rise in it, and wet it considerably above the surface of that into which you dip it.

EMILY.

In making tea I have often observed that effect, without being able to account for it.

MRS. B.

Now that you are acquainted with the attraction of cohesion, I must endeavour to explain to you that of *Gravitation*, which is a modification of the same power; the first is perceptible only in very minute particles, and at very small distances; the other acts on the largest bodies, and extends to immense distances.

### EMILY.

You astonish me : surely you do not mean to say, that large bodies attract each other.

### MRS. B.

Indeed I do : let us take, for example, one of the largest bodies in nature, and observe whether it does not attract other bodies. What is it that occasions the fall of this book, when I no longer support it?

### EMILY.

Can it be the attraction of the earth? I thought that all bodies had a natural tendency to fall.

### MRS. B.

They have a natural tendency to fall, it is true; but that tendency is produced entirely by the attraction of the earth : the earth being so much larger than any body on its surface, forces every body, which is not supported, to fall upon it.

### EMILY.

If the tendency which bodies have to fall results from the earth's attractive power, the earth itself can have no such tendency, since it cannot attract itself, and therefore it requires no support to prevent it from falling. Yet the idea that bodies do

not fall of their own accord, but that they are
drawn towards the earth by its attraction, is so new
and strange to me, that I know not how to reconcile myself to it.

MRS. B.

When you are accustomed to consider the fall of
bodies as depending on this cause, it will appear to
you as natural, and surely much more satisfactory,
than if the cause of their tendency to fall were totally
unknown.   Thus you understand, that all matter is
attractive, from the smallest particle to the largest
mass; and that bodies attract each other with a
force proportional to the quantity of matter they
contain.

EMILY.

I do not perceive any difference between the attraction of cohesion and that of gravitation: is it
not because every particle of matter is endowed
with an attractive power, that large bodies, consisting of a great number of particles, are so strongly
attractive?

MRS. B.

True.   There is, however, this difference between the attraction of particles and that of masses,
that the former is stronger than the latter, in proportion to the quantity of matter.   Of this you have
an instance in the attraction of capillary tubes, in

which liquids ascend by the attraction of cohesion, in opposition to that of gravity. It is on this account that it is necessary that the bore of the tube should be extremely small; for if the column of water within the tube is not very minute, the attraction would not be able either to raise or support its weight, in opposition to that of gravity.

You may observe, also, that all solid bodies are enabled by the force of the cohesive attraction of their particles to resist that of gravity, which would otherwise disunite them, and bring them to a level with the ground, as it does in the case of liquids, the cohesive attraction of which is not sufficient to enable it to resist the power of gravity.

#### EMILY.

And some solid bodies appear to be of this nature, as sand and powder for instance; there is no attraction of cohesion between their particles?

#### MRS. B.

Every grain of powder or sand is composed of a great number of other more minute particles, firmly united by the attraction of cohesion; but amongst the separate grains there is no sensible attraction, because they are not in sufficiently close contact.

#### EMILY.

Yet they actually touch each other?

MRS. B.

The surface of bodies is in general so rough and uneven, that when in actual contact, they touch each other only by a few points. Thus, if I lay upon the table this book, the binding of which appears perfectly smooth, yet so few of the particles of its under surface come in contact with the table, that no sensible degree of cohesive attraction takes place; for you see, that it does not stick, or cohere to the table, and I find no difficulty in lifting it off.

It is only when surfaces perfectly flat and well polished are placed in contact, that the particles approach in sufficient number, and closely enough, to produce a sensible degree of cohesive attraction. Here are two hemispheres of polished metal, I press their flat surfaces together, having previously interposed a few drops of oil, to fill up every little porous vacancy. Now try to separate them.

EMILY.

It requires an effort beyond my strength, though there are handles for the purpose of pulling them asunder. Is the firm adhesion of the two hemispheres, merely owing to the attraction of cohesion?

MRS. B.

There is no force more powerful, since it is by this, that the particles of the hardest bodies are

C

held together.   It would require a weight of several pounds, to separate these hemispheres.

In making a kaleidoscope, I recollect that the two plates of glass, which were to serve as mirrors, stuck so fast together, that I imagined some of the gum I had been using had by chance been interposed between them; but now I make no doubt but that it was their own natural cohesive attraction which produced this effect.

Very probably it was so; for plate-glass has an extremely smooth flat surface, admitting of the contact of a great number of particles, between two plates, laid one over the other.

But, Mrs. B., the cohesive attraction of some bodies is much greater than that of others; thus glue, gum, and paste, cohere with singular tenacity.

That is owing to the peculiar chemical properties of those bodies, independently of their cohesive attraction.

There are some other kinds or modifications of

attraction peculiar to certain bodies; namely, that
of magnetism, and of electricity; but we shall confine our attention merely to the attraction of
cohesion and of gravity; the examination of the
latter we shall resume at our next meeting.

# CONVERSATION II.

## ON THE ATTRACTION OF GRAVITY.

ATTRACTION OF GRAVITATION, CONTINUED. — OF
WEIGHT. — OF THE FALL OF BODIES. — OF THE
RESISTANCE OF THE AIR. — OF THE ASCENT OF
LIGHT BODIES.

#### EMILY.

I HAVE related to my sister Caroline all that you
have taught me of natural philosophy, and she has
been so much delighted by it, that she hopes you
will have the goodness to admit her to your lessons.

#### MRS. B.

Very willingly; but I did not think you had any
taste for studies of this nature, Caroline?

#### CAROLINE.

I confess, Mrs. B., that hitherto I had formed no
very agreeable idea, either of philosophy, or phi-

losophers; but what Emily has told me, has excited my curiosity so much, that I shall be highly pleased if you will allow me to become one of your pupils.

### MRS. B.

I fear that I shall not find you so tractable a scholar as Emily; I know that you are much biassed in favour of your own opinions.

### CAROLINE.

Then you will have the greater merit in reforming them, Mrs. B.; and after all the wonders that Emily has related to me, I think I stand but little chance against you and your attractions.

### MRS. B.

You will, I doubt not, advance a number of objections; but these I shall willingly admit, as they will be a means of elucidating the subject. Emily, do you recollect the names of the general properties of bodies?

### EMILY.

Impenetrability, extension, figure, divisibility, inertia, and attraction.

### MRS. B.

Very well. You must remember that these are properties common to all bodies, and of which they cannot be deprived; all other properties of bodies

c 3

are called accidental, because they depend on the relation or connection of one body to another.

### CAROLINE.

Yet surely, Mrs. B., there are other properties which are essential to bodies, besides those you have enumerated. Colour and weight, for instance, are common to all bodies, and do not arise from their connection with each other, but exist in the bodies themselves; these, therefore, cannot be accidental qualities?

### MRS. B.

I beg your pardon; these properties do not exist in bodies independently of their connection with other bodies.

### CAROLINE.

What! have bodies no weight? Does not this table weigh heavier than this book; and, if one thing weighs heavier than another, must there not be such a thing as weight?

### MRS. B.

No doubt: but this property does not appear to be essential to bodies; it depends upon their connection with each other. Weight is an effect of the power of attraction, without which the table and the book would have no weight whatever.

EMILY.

I think I understand you; is it not the attraction
of gravity, which makes bodies heavy?

MRS. B.

You are right. I told you that the attraction of
gravity was proportioned to the quantity of matter
which bodies contained: now the earth consisting
of a much greater quantity of matter than any body
upon its surface, the force of its attraction must
necessarily be greatest, and must draw every thing
towards it; in consequence of which, bodies that are
unsupported fall to the ground, whilst those that
are supported press upon the object which prevents
their fall, with a weight equal to the force with
which they gravitate towards the earth.

CAROLINE.

The same cause then which occasions the fall of
bodies, produces also their weight. It was very
dull in me not to understand this before, as it is
the natural and necessary consequence of attraction; but the idea that bodies were not really
heavy of themselves, appeared to me quite incomprehensible. But, Mrs. B., if attraction is a property essential to matter, weight must be so likewise; for how can one exist without the other?

MRS. B.

Suppose there were but one body existing in universal space, what would its weight be?

CAROLINE.

That would depend upon its size; or, more accurately speaking, upon the quantity of matter it contained.

EMILY.

No, no; the body would have no weight, whatever were its size; because nothing would attract it. Am I not right, Mrs. B.?

MRS. B.

You are: you must allow, therefore, that it would be possible for attraction to exist without weight; for each of the particles of which the body was composed, would possess the power of attraction; but they could exert it only amongst themselves; the whole mass, having nothing to attract, or to be attracted by, would have no weight.

CAROLINE.

I am now well satisfied that weight is not essential to the existence of bodies; but what have you to object to colours, Mrs. B.; you will not, I think, deny that they really exist in the bodies themselves.

#### MRS. B.

When we come to treat of the subject of colours, I trust that I shall be able to convince you, that colours are likewise accidental qualities, quite distinct from the bodies to which they appear to belong.

#### CAROLINE.

Oh do pray explain it to us now, I am so very curious to know how that is possible.

#### MRS. B.

Unless we proceed with some degree of order and method, you will in the end find yourself but little the wiser for all you learn.   Let us therefore go on regularly, and make ourselves well acquainted with the general properties of bodies, before we proceed further.

#### EMILY.

To return, then, to attraction, (which appears to me by far the most interesting of them, since it belongs equally to all kinds of matter,) it must be mutual between two bodies; and if so, when a stone falls to the earth, the earth should rise part of the way to meet the stone?

#### MRS. B.

Certainly; but you must recollect that the force of attraction is proportioned to the quantity of matter which bodies contain, and if you consider

the difference there is in that respect, between a
stone and the earth; you will not be surprised that
you do not perceive the earth rise to meet the stone;
for though it is true that a mutual attraction takes
place between the earth and the stone, that of the
latter is so very small in comparison to that of the
former, as to render its effect insensible.

EMILY.

But since attraction is proportioned to the quantity of matter which bodies contain, why do not the
hills attract the houses and churches towards them?

CAROLINE.

Heavens, Emily, what an idea! How can the
houses and churches be moved, when they are so
firmly fixed in the ground?

MRS. B.

Emily's question is not absurd, and your answer,
Caroline, is perfectly just; but can you tell us why
the houses and churches are so firmly fixed in the
ground?

CAROLINE.

I am afraid I have answered right by mere chance;
for I begin to suspect that bricklayers and carpenters
could give but little stability to their buildings, without the aid of attraction.

### MRS. B.

It is certainly the cohesive attraction between the bricks and the mortar, which enables them to build walls, and these are so strongly attracted by the earth, as to resist every other impulse; otherwise they would necessarily move towards the hills and the mountains; but the lesser force must yield to the greater. There are, however, some circumstances in which the attraction of a large body has sensibly counteracted that of the earth. If, whilst standing on the declivity of a mountain, you hold a plumb-line in your hand, the weight will not fall perpendicular to the earth, but incline a little towards the mountain; and this is owing to the lateral, or sideways attraction of the mountain, interfering with the perpendicular attraction of the earth.

### EMILY.

But the size of a mountain is very trifling, compared to the whole earth?

### MRS. B.

Attraction, you must recollect, diminishes with distance; and in the example of the plumb-line, the weight suspended is considerably nearer to the mountain than to the centre of the earth.

### CAROLINE.

Pray, Mrs. B., do the two scales of a balance hang parallel to each other?

MRS. B.

You mean, I suppose, in other words to inquire whether two lines which are perpendicular to the earth, are parallel to each other? I believe I guess the reason of your question; but I wish you would endeavour to answer it without my assistance.

CAROLINE.

I was thinking that such lines must both tend by gravity to the same point, the centre of the earth; now lines tending to the same point cannot be parallel, as parallel lines are always at an equal distance from each other, and would never meet.

MRS. B.

Very well explained: you see now the use of your knowledge of parallel lines; had you been ignorant of their properties, you could not have drawn such a conclusion. This may enable you to form an idea of the great advantage to be derived even from a slight knowledge of geometry, in the study of natural philosophy; and if, after I have made you acquainted with the first elements, you should be tempted to pursue the study, I would advise you to prepare yourselves by acquiring some knowledge of geometry. This science would teach you that lines which fall perpendicular to the surface of a sphere cannot be parallel, because they would all meet, if prolonged to the centre of the sphere; while lines that fall perpendicular to a plane or flat surface, are

PLATE I.

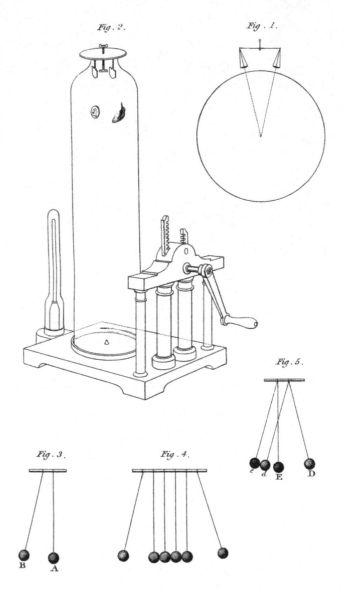

Fig. 2.

Fig. 1.

Fig. 5.

Fig. 3.

Fig. 4.

Published by Longman & Co. June 1ˢᵗ 1840.

Lowry Sc.

always parallel, because, if prolonged, they would never meet.

EMILY.

And yet a pair of scales, hanging perpendicular to the earth, appear parallel?

MRS. B.

Because the sphere is so large, and the scales consequently converge so little, that their inclination is not perceptible to our senses; if we could construct a pair of scales whose beam would extend several degrees, their convergence would be very obvious; but as this cannot be accomplished, let us draw a small figure of the earth, and then we may make a pair of scales of the proportion we please. (fig. 1. Plate I.)

CAROLINE.

This figure renders it very clear : then two bodies cannot fall to the earth in parallel lines?

MRS. B.

Never.

CAROLINE.

The reason that a heavy body falls quicker than a light one, is, I suppose, because the earth attracts it more strongly?

MRS. B.

The earth, it is true, attracts a heavy body more than a light one; but that would not make the one fall quicker than the other.

### CAROLINE.

Yet, since it is attraction that occasions the fall of bodies, surely the more a body is attracted, the more rapidly it will fall. Besides, experience proves it to be so. Do we not every day see heavy bodies fall quickly, and light bodies slowly.

### EMILY.

It strikes me, as it does Caroline, that as attraction is proportioned to the quantity of matter, the earth must necessarily attract a body which contains a great quantity more strongly, and therefore bring it to the ground sooner than one consisting of a smaller quantity.

### MRS. B.

You must consider, that if heavy bodies are attracted more strongly than light ones, they require more attraction to make them fall. Remember that bodies have no natural tendency to fall, any more than to rise, or to move laterally, and that they will not fall unless impelled by some force; now this force must be proportioned to the quantity of matter it has to move: a body consisting of 1000 particles of matter, for instance, requires ten times as much attraction to bring it to the ground in the same space of time as a body consisting of only 100 particles.

CAROLINE.

I do not understand that; for it seems to me, that the heavier a body is, the more easily and readily it falls.

EMILY.

I think I now comprehend it; let me try if I can explain it to Caroline. Suppose that I draw towards me two weighty bodies, the one of 100lbs. the other of 1000lbs., must I not exert ten times as much strength to draw the larger one to me, in the same space of time as is required for the smaller one? And if the earth draws a body of 1000lbs. weight to it in the same space of time that it draws a body of 100lbs., does it not follow that it attracts the body of 1000lbs. weight with ten times the force that it does that of 100lbs.?

CAROLINE.

I comprehend your reasoning perfectly; but if it were so, the body of 1000lbs. weight, and that of 100lbs. would fall with the same rapidity; and the consequence would be, that all bodies, whether light or heavy, being at an equal distance from the ground, would fall to it in the same space of time: now it is very evident that this conclusion is absurd; experience every instant contradicts it: observe how much sooner this book reaches the floor than this sheet of paper, when I let them drop together.

### EMILY.

That is an objection I cannot answer. I must refer it to you Mrs. B.

### MRS. B.

I trust that we shall not find it insurmountable. It is true that, according to the laws of attraction, all bodies at an equal distance from the earth, should fall to it in the same space of time; and this would actually take place if no obstacle intervened to impede their fall. But bodies fall through the air, and it is the resistance of the air which makes bodies of different density fall with different degrees of velocity. They must all force their way through the air, but dense heavy bodies overcome this obstacle more easily than rarer and lighter ones.

The resistance which the air opposes to the fall of bodies is proportioned to their surface, not to their weight; the air being inert, cannot exert a greater force to support the weight of a cannon-ball, than it does to support the weight of a ball (of the same size) made of leather; but the cannon-ball will overcome this resistance more easily, and fall to the ground, consequently, quicker than the leather ball.

### CAROLINE.

This is very clear, and solves the difficulty perfectly. The air offers the same resistance to a bit of lead and a bit of feather of the same size; yet the one seems to meet with no obstruction in its fall, whilst the other is evidently resisted and supported for some time by the air.

EMILY.

The larger the surface of a body, then, the more air it covers, and the greater is the resistance it meets with from it.

MRS. B.

Certainly; observe the manner in which this sheet of paper falls; it floats awhile in the air, and then gently descends to the ground.   I will roll the same piece of paper up into a ball: it offers now but a small surface to the air, and encounters there-fore but little resistance: see how much more rapidly it falls.

The heaviest bodies may be made to float awhile in the air, by making the extent of their surface counterbalance their weight.   Here is some gold, which is the most dense body we are acquainted with, but it has been beaten into a very thin leaf, and offers so great an extent of surface in propor-tion to its weight, that its fall, you see, is still more retarded by the resistance of the air than that of the sheet of paper.

CAROLINE.

That is very curious; and it is, I suppose, upon

the same principle that iron boats may be made to float on water?

But, Mrs. B., if the air is a real body, is it not also subjected to the laws of gravity?

MRS. B.

Undoubtedly.

CAROLINE.

Then why does it not, like all other bodies, fall to the ground?

MRS. B.

On account of its spring or elasticity. The air is an *elastic fluid*; a species of bodies, the peculiar property of which is to resume, after compression, their original dimensions; and you must consider the air of which the atmosphere is composed as existing in a state of compression, for its particles being drawn towards the earth by gravity, are brought closer together than they would otherwise be, but the spring or elasticity of the air by which it endeavours to resist compression, gives it a constant tendency to expand itself, so as to resume the dimensions it would naturally have, if not under the influence of gravity. The air may therefore be said constantly to struggle with the power of gravity without being able to overcome it. Gravity thus confines the air to the regions of our globe, whilst its elasticity prevents it from falling like other bodies to the ground.

EMILY.

The air then is, I suppose, thicker, or I should rather say more dense, near the surface of the earth, than in the higher regions of the atmosphere; for that part of the air which is nearer the surface of the earth, must be most strongly attracted.

MRS. B.

The diminution of the force of gravity, at so small a distance as that to which the atmosphere extends (compared with the size of the earth) is so inconsiderable as to be scarcely sensible; but the pressure of the upper parts of the atmosphere on those beneath, renders the air near the surface of the earth much more dense than the upper regions. The pressure of the atmosphere has been compared to that of a pile of fleeces of wool, in which the lower fleeces are pressed together by the weight of those above; these lie light and loose, in proportion as they approach the uppermost fleece, which receives no external pressure, and is confined merely by the force of its own gravity.

CAROLINE.

It has just occurred to me that there are some bodies which do not gravitate towards the earth. Smoke and steam, for instance, rise instead of falling.

MRS. B.

It is still gravity which produces their ascent; at least, were that power destroyed, these bodies would not rise.

CAROLINE.

I shall be out of conceit with gravity, if it is so inconsistent in its operations.

MRS. B.

There is no difficulty in reconciling this apparent inconsistency of effect. The air near the earth is heavier than smoke, steam, or other vapours; it consequently not only supports these light bodies, but forces them to rise, till they reach a part of the atmosphere, the weight of which is not greater than their own, and then they remain stationary. Look at this bason of water; why does the piece of paper which I throw into it float on the surface?

EMILY.

Because, being lighter than the water, it is supported by it.

MRS. B.

And now that I pour more water into the bason, why does the paper rise?

EMILY.

The water being heavier than the paper, gets beneath it, and obliges it to rise.

MRS. B.

In a similar manner are smoke and vapour forced upwards by the air; but these bodies do not, like the paper, ascend to the surface of the fluid, because, as we observed before, the air being thinner and lighter as it is more distant from the earth, vapours rise only till they attain a region of air of their own density.   Smoke, indeed, ascends but a very little way; it consists of minute particles of fuel carried up by a current of heated air from the fire below: heat, you recollect, expands all bodies; it consequently rarefies air, and renders it lighter than the colder air of the atmosphere; the heated air from the fire carries up with it vapour and small particles of the combustible materials which are burning in the fire.   When this current of hot air is cooled by mixing with that of the atmosphere, the minute particles of coal or other combustible fall, and it is this which produces the small black flakes which render the air and every thing in contact with it, in London, so dirty.

CAROLINE.

You must, however, allow me to make one more objection to the universal gravity of bodies; which is the ascent of air balloons, the materials of which are undoubtedly heavier than air: how, therefore, can they be supported by it?

### MRS. B.

I admit that the materials of which balloons are made are heavier than the air; but the air with which they are filled is an elastic fluid, of a different nature from the atmospheric air, and considerably lighter; so that, on the whole, the balloon is lighter than the air which it displaces, and consequently will rise, on the same principle as smoke and vapour. Now Emily, let me hear if you can explain how the gravity of bodies is modified by the effect of the air?

### EMILY.

The air forces bodies which are lighter than itself to ascend; those that are of an equal weight will remain stationary in it; and those that are heavier will descend through it: but the air will have some effect on these last; for if they are not much heavier, they will with difficulty overcome the resistance they meet with in passing through it, they will be borne up by it, and their fall will be more or less retarded.

### MRS. B.

Very well. Observe how slowly this light feather falls to the ground, while a heavier body, like this marble, overcomes the resistance which the air makes to its descent much more easily, and its fall is proportionally more rapid. I now throw a pebble into this tub of water; it does not reach the bot-

tom near so soon as if there were no water in the
tub, because it meets with resistance from the water.   Suppose that we could empty the tub, not only
of water, but of air also, the pebble would then fall
quicker still, as it would in that case meet with no
resistance at all to counteract its gravity.

Thus you see that it is not the different degrees
of gravity, but the resistance of the air, which, prevents bodies of different weight from falling with
equal velocities; if the air did not bear up the feather, it would reach the ground as soon as the
marble.

CAROLINE.

I make no doubt that it is so; and yet I do not
feel quite satisfied.   I wish there was any place void
of air, in which the experiment could be made.

MRS. B.

If that proof will satisfy your doubts, I can give it
you.   Here is a machine called an *air pump*, (fig. 2.
pl. I.) by means of which the air may be expelled from
any close vessel which is placed over this opening,
through which the air is pumped out.   Glasses of
various shapes, usually called *receivers*, are employed
for this purpose.   We shall now exhaust the air
from this tall receiver which is placed over the opening, and we shall find that bodies of whatever

weight or size within it, will fall from the top to the
bottom in the same space of time.

CAROLINE.

Oh, I shall be delighted with this experiment!
what a curious machine! how can you put the two
bodies of different weight within the glass, without
admitting the air.

MRS. B.

A guinea and a feather are already placed there for
the purpose of the experiment: here is you see a
contrivance to fasten them in the upper part of the
glass; as soon as the air is pumped out, I shall turn
this little screw, by which means the brass plates
which support them will be inclined, and the two
bodies will fall.— Now I believe I have pretty well
exhausted the air.

CAROLINE.

Pray let me turn the screw.—I declare, they both
reached the bottom at the same instant! Did you
see, Emily, the feather appeared as heavy as the
guinea?

EMILY.

Exactly; and fell just as quickly. How wonderful
this is! what a number of entertaining experiments
might be made with this machine!

#### MRS. B.

No doubt there are a great variety; but we shall reserve them to elucidate the subjects to which they relate: if I had not explained to you why the guinea and the feather fell with equal velocity, you would not have been so well pleased with the experiment.

#### EMILY.

I should have been as much surprised, but not so much interested; besides, experiments help to imprint on the memory the facts they are intended to illustrate; it will be better therefore for us to restrain our curiosity, and wait for other experiments in their proper places.

#### CAROLINE.

Pray by what means is the air exhausted in this receiver?

#### MRS. B.

You must learn something of mechanics in order to understand the construction of a pump. At our next meeting, therefore, I shall endeavour to make you acquainted with the laws of motion, as an introduction to that subject.

# CONVERSATION III,

## ON THE LAWS OF MOTION.

MRS. B.

THE science of mechanics is founded on the laws
of motion; it will therefore be necessary to make
you acquainted with these laws before we examine
the mechanical powers. Tell me, Caroline, what
do you understand by the word motion?

CAROLINE.

I think I understand it perfectly, though I am at

a loss to describe it. Motion is the act of moving about, going from one place to another, it is the contrary of remaining at rest.

MRS. B.

Very well. Motion then consists in a change of place; a body is in motion whenever it is changing its situation with regard to a fixed point.

Now since we have observed that one of the general properties of bodies is Inertia, that is, an entire passiveness either with regard to motion or rest, it follows that a body cannot move without being put into motion; the power which puts a body into motion is called *force*; thus the stroke of the hammer is the force which drives the nail; the pulling of the horse that which draws the carriage, &c. Force then is the cause which produces motion.

EMILY.

And may we not say that gravity is the force which occasions the fall of bodies?

MRS. B.

Undoubtedly. I had given you the most familiar illustrations in order to render the explanation clear; but since you seek for more scientific examples, you may say that cohesion is the force which binds the particles of bodies together, and heat that which drives them asunder.

D 2

The motion of a body acted upon by a single force is always in a straight line, in the direction in which it received the impulse.

CAROLINE.

That is very natural; for as the body is inert, and can move only because it is impelled, it will move only in the direction in which it is impelled. The degree of quickness with which it moves, must, I suppose, also depend upon the degree of force with which it is impelled.

MRS. B.

Yes; the rate at which a body moves, or the shortness of the time which it takes to move from one place to another, is called its velocity; and it is one of the laws of motion that the velocity of the moving body is proportional to the force by which it is put in motion. We must distinguish between absolute and relative velocity.

The velocity of a body is called *absolute*, if we consider the motion of the body in space, without any reference to that of other bodies. When for instance a horse goes fifty miles in ten hours, his velocity is five miles an hour.

The velocity of a body is termed *relative*, when compared with that of another body which is itself in motion. For instance, if one man walks at the rate of a mile an hour, and another at the rate of two miles an hour, the relative velocity of the latter is double that of the former; but the absolute velocity of the one is one mile, and that of the other two miles an hour.

EMILY.

Let me see if I understand it—The relative velocity of a body is the degree of rapidity of its motion compared with that of another body; thus if one ship sail three times as far as another ship in the same space of time, the velocity of the former is equal to three times that of the latter.

MRS. B.

The general rule may be expressed thus: the velocity of a body is measured by the space over which it moves, divided by the time which it employs in that motion: thus if you travel one hundred miles in twenty hours, what is your velocity in each hour?

EMILY.

I must divide the space, which is one hundred miles, by the time, which is twenty hours, and the answer will be five miles an hour. Then, Mrs. B., may we not reverse this rule and say, that the time is equal to the space divided by the velocity; since the space one hundred miles, divided by the velocity five miles, gives twenty hours for the time?

MRS. B.

Certainly; and we may say also that space is equal to the velocity multiplied by the time.  Can you tell me, Caroline, how many miles you will have travelled, if your velocity is three miles an hour and you travel six hours?

CAROLINE.

Eighteen miles; for the product of 3 multiplied by 6, is 18.

MRS. B.

I suppose that you understand what is meant by the terms *uniform, accelerated* and *retarded* motion.

EMILY.

I conceive uniform motion to be that of a body whose motion is regular, and at an equal rate throughout; for instance, a horse that goes an equal number of miles every hour.  But the hand of a watch is a much better example, as its motion is so regular as to indicate the time.

MRS. B.

You have a right idea of uniform motion; but it would be more correctly expressed by saying, that the motion of a body is uniform when it passes over equal spaces in equal times.  Uniform motion is produced by a force having acted on a body once, and having ceased to act; as for instance, the stroke of a bat on a cricket ball.

CAROLINE.

But the motion of a cricket ball is not uniform; its velocity gradually diminishes till it falls to the ground.

MRS. B.

Recollect that the cricket ball is inert, and has no more power to stop than to put itself in motion; if it falls, therefore, it must be stopped by some force superior to that by which it was projected, and which destroys its motion.

CAROLINE.

And it is no doubt the force of gravity which counteracts and destroys that of projection; but if there were no such power as gravity, would the cricket ball never stop?

MRS. B.

If neither gravity nor any other force, such as the resistance of the air, opposed its motion, the cricket ball, or even a stone thrown by the hand, would proceed onwards in a right line, and with an uniform velocity for ever.

CAROLINE.

You astonish me! I thought that it was impossible to produce perpetual motion?

Perpetual motion cannot be produced by art, because gravity ultimately destroys all motion that human powers can produce.

But independently of the force of gravity, I cannot conceive that the little motion I am capable of giving to a stone would put it in motion for ever.

The quantity of motion you communicated to the stone would not influence its duration; if you threw it with little force it would move slowly, for its velocity, you must remember, will be proportional to the force with which it is projected; but if there is nothing to obstruct its passage, it will continue to move with the same velocity, and in the same direction as when you first projected it.

This appears to me quite incomprehensible; we do not meet with a single instance of it in nature.

I beg your pardon. When you come to study the motion of the celestial bodies, you will find that nature abounds with examples of perpetual motion; and that it conduces as much to the harmony of the system of the universe, as the prevalence of it would to the destruction of all comfort on our globe. The wisdom of Providence has therefore ordained insurmountable obstacles to perpetual motion here below, and though these obstacles often compel us to contend with great difficulties, yet there results from it that order, regularity and repose, so essential to the preservation of all the various beings of which this world is composed.

Now can you tell me what is *retarded motion?*

### CAROLINE.

Retarded motion is that of a body which moves every moment slower and slower: thus when I am tired with walking fast, I slacken my pace; or when a stone is thrown upwards, its velocity is gradually diminished by the power of gravity.

### MRS. B.

Retarded motion is produced by some force acting upon the body in a direction opposite to that which first put it in motion: you who are an animated being, endowed with power and will, may slacken your pace, or stop to rest when you are tired; but inert matter is incapable of any feeling of fatigue, can never slacken its pace, and never stop, unless retarded or arrested in its course by some opposing force; and as it is the laws of inert bodies which mechanics treats of, I prefer your

illustration of the stone retarded in its ascent. Now
Emily, it is your turn; what is *accelerated motion?*

EMILY.

Accelerated motion, I suppose, takes place when
the velocity of a body is increased; if you had not
objected to our giving such active bodies as ourselves as examples, I should say that my motion is
accelerated if I change my pace from walking to
running. I cannot think of any instance of accelerated motion in inanimate bodies; all motion of
inert matter seems to be retarded by gravity.

MRS. B.

Not in all cases; for the power of gravitation
sometimes produces accelerated motion; for instance,
a stone falling from a height moves with a regularly
accelerated motion.

EMILY.

True; because the nearer it approaches the earth,
the more it is attracted by it.

MRS. B.

You have mistaken the cause of its acceleration
of motion; for though it is true that the force of
gravity increases as a body approaches the earth,
the difference is so trifling at any small distance
from its surface as not to be perceptible.

Accelerated motion is produced when the force

which put a body in motion continues to act upon
it during its motion, so that its motion is continually
increased.    When a stone falls from a height, the
impulse which it receives from gravity during the
first instant of its fall, would be sufficient to bring
it to the ground with a uniform velocity : for, as we
have observed, a body having been once acted upon
by a force, will continue to move with a uniform
velocity; but the stone is not acted upon by gravity
merely at the first instant of its fall, this power continues to impel it during the whole of its descent,
and it is this continued impulse which accelerates
its motion.

EMILY.

I do not quite understand that.

MRS. B.

Let us suppose that the instant after you have let
fall a stone from a high tower, the force of gravity
were annihilated, the body would nevertheless continue to move downwards, for it would have received
a first impulse from gravity, and a body once put
in motion will not stop unless it meets with some
obstacle to impede its course; in this case its velocity would be uniform, for though there would be
no obstacle to obstruct its descent, there would
no force to accelerate it.

EMILY.

That is very clear.

D 6

Then you have only to add the power of gravity constantly acting on the stone during its descent, and it will not be difficult to understand that its motion will become accelerated, since the gravity which acts on the stone during the first instant of its descent, will continue in force every instant till it reaches the ground. Let us suppose that the impulse given by gravity to the stone during the first instant of its descent be equal to one, the next instant we shall find that an additional impulse gives the stone an additional velocity equal to one, so that the accumulated velocity is now equal to two; the following instant another impulse increases the velocity to three, and so on till the stone reaches the ground.

CAROLINE.

Now I understand it; the effects of preceding impulses must be added to the subsequent velocities.

MRS. B.

Yes; it has been ascertained, both by experiment and calculations, which it would be too difficult for us to enter into, that heavy bodies descending from a height by the force of gravity, fall sixteen feet the first second of time, three times that distance in the next, five times in the third second, seven times in the fourth, and so on, regularly increasing their velocities according to the number of seconds during which the body has been falling.

EMILY.

If you throw a stone perpendicularly upwards, is it not the same length of time ascending that it is descending?

MRS. B.

Exactly; in ascending, the velocity is diminished by the force of gravity; in descending, it is accelerated by it.

CAROLINE.

I should then have imagined that it would have fallen quicker than it rose?

MRS. B.

You must recollect that the force with which it is projected must be taken into the account; and that this force is overcome and destroyed by gravity before the body falls.

CAROLINE.

But the force of projection given to a stone in throwing it upwards, cannot always be equal to the force of gravity in bringing it down again, for the force of gravity is always the same, whilst the degree of impulse given to the stone is optional; I may throw it up gently, or with violence.

MRS. B.

If you throw it gently, it will not rise high; perhaps only sixteen feet, in which case it will fall in one second of time. Now it is proved by experiment, that an impulse requisite to project a body sixteen feet upwards, will make it ascend that height in one second; here then the times of the ascent and descent are equal. But supposing it be required to throw a stone twice that height, the force must be proportionally greater.

MRS. B.

You see then, that the impulse of projection in throwing a body upwards, is always equal to the action of the force of gravity during its descent; and that it is the greater or less distance to which the body rises, that makes these two forces balance each other.

I must now explain to you what is meant by the *momentum* of bodies. It is the force, or power, with which a body in motion, strikes against another body. The momentum of a body is composed of its *quantity of matter*, multiplied by its *quantity of motion ;* in other words, its weight and its velocity.

CAROLINE.

The quicker a body moves, the greater, no doubt, must be the force with which it would strike against another body.

EMILY.

Therefore a small body may have a greater momentum than a large one, provided its velocity be sufficiently greater; for instance, the momentum of an arrow shot from a bow, must be greater than a stone thrown by the hand.

### CAROLINE.

We know also by experience, that the heavier a body is, the greater is its force; it is not therefore difficult to understand, that the whole power or momentum of a body must be composed of these two properties: but I do not understand, why they should be *multiplied*, the one by the other; I should have supposed that the quantity of matter should have been *added* to the quantity of motion?

### MRS. B.

It is found by experiment, that if the weight of a body is represented by the number 3, and its velocity also by 3, its momentum will be represented by 9; not 6, as would be the case, were these figures added, instead of being multiplied together. I recommend it to you to be careful to remember the definition of the momentum of bodies, as it is one of the most important points in mechanics; you will find, that it is from opposing motion to matter, that machines derive their powers. *

---

* In comparing together the momenta of different bodies, we must be attentive to measure their weights and velocities,

The *reaction* of bodies, is the next law of motion which I must explain to you.    When a body in motion strikes against another body, it meets with resistance from it; the resistance of the body at rest, will be equal to the blow struck by the body in motion; or to express myself in philosophical language, *action* and *re-action* will be equal, and in opposite directions.

CAROLINE.

Do you mean to say, that the action of the body which strikes, is returned with equal force by the body which receives the blow.

MRS. B.

Exactly.

CAROLINE.

But if a man strikes another on the face with his fist, he surely does not receive as much pain by the re-action, as he inflicts by the blow?

---

by the same denomination of weights and of spaces, otherwise the results would not agree.    Thus if we estimate the weight of one body in ounces, we must estimate the weight of the rest also in ounces, and not in pounds; and in computing the velocities, in like manner we should adhere to the same standard of measure, both of space and of time; as for instance, the number of feet in one second, or of miles in one hour.

MRS. B.

No; but this is simply owing to the knuckles having much less feeling, than the face.

Here are two ivory balls suspended by threads, (Plate I. fig. 3.) draw one of them, A, a little on one side, — now let it go; — it strikes, you see, against the other ball B, and drives it off, to a distance equal to that through which the first ball fell; but the motion of A is stopped, because when it struck B, it received in return a blow equal to that it gave, and its motion was consequently destroyed.

EMILY.

I should have supposed, that the motion of the ball A was destroyed, because it had communicated all its motion to B.

MRS. B.

It is perfectly true, that when one body strikes against another, the quantity of motion communicated to the second body, is lost by the first; but this loss proceeds from the action of the body which is struck.

Here are six ivory balls hanging in a row, (fig. 4.) draw the first out of the perpendicular, and let it fall against the second. None of the balls appear to move, you see, except the last, which flies off as far as the first ball fell; can you explain this?

CAROLINE.

I believe so. When the first ball struck the second, it received a blow in return, which destroyed its motion; the second ball, though it did not appear to move, must have struck against the third; the re-action of which set it at rest; the action of the third ball must have been destroyed by the re-action of the fourth, and so on, till motion was communicated to the last ball, which, not being re-acted upon, flies off.

MRS. B.

Very well explained. Observe, that it is only when bodies are elastic, as these ivory balls are, that the stroke returned is equal to the stroke given. I will shew you the difference with these two balls of clay, (fig. 5.) which are not elastic; when you raise one of these, D, out of the perpendicular, and let it fall against the other, E, the re-action of the latter, on account of its not being elastic, is not sufficient to destroy the motion of the former; only part of the motion of D will be communicated to E, and the two balls will move on together to d and e, which is not to so great a distance as that through which D fell.

Observe how useful re-action is in nature. Birds in flying, strike the air with their wings, and it is the re-action of the air which enables them to rise, or advance forwards; re-action being always in a contrary direction to action.

CAROLINE.

I thought that birds might be lighter than the air, when their wings were expanded, and by that means enabled to fly.

MRS. B.

When their wings are spread, they are better supported by the air, as they cover a greater extent of surface; but they are still much too heavy to remain in that situation, without continually flapping their wings, as you may have noticed, when birds hover over their nests; the force with which their wings strike against the air must equal the weight of their bodies, in order that the re-action of the air may be able to support that weight; the bird will then remain stationary. If the stroke of the wings is greater than is required merely to support the bird, the re-action of the air will make it rise; if it be less, it will gently descend; and you may have observed the lark, sometimes remaining with its wings extended, but motionless; in this state it drops rapidly into its nest.

CAROLINE.

What a beautiful effect this is of the law of re-action! But if flying is merely a mechanical

operation, Mrs. B., why should we not construct
wings, adapted to the size of our bodies, fasten them
to our shoulders, move them with our arms, and
soar into the air.

MRS. B.

Such an experiment has been repeatedly at-
tempted, but never with success; and it is now
considered as totally impracticable. The muscular
power of birds is greater in proportion to their
weight than that of man; were we therefore furnished with wings sufficiently large to enable us to
fly, we should not have strength to put them in
motion.

In swimming, a similar action is produced on
the water, as that on the air in flying; and also in
rowing; you strike the water with the oars, in a
direction opposite to that in which the boat is required to move, and it is the re-action of the water
on the oars which drives the boat along.

EMILY.

You said, that it was in elastic bodies only, that
re-action was equal to action; pray what bodies
are elastic besides the air?

MRS. B.

In speaking of the air, I think we defined elasticity to be a property, by means of which bodies

14

that are compressed return to their former state. If I bend this cane, as soon as I leave it at liberty it recovers its former position; if I press my finger upon your arm, as soon as I remove it, the flesh, by virtue of its elasticity, rises and destroys the impression I made. Of all bodies, the air is the most eminent for this property, and it has thence obtained the name of elastic fluid. Hard bodies are in the next degree elastic; if two ivory, or metallic balls are struck together, the parts at which they touch will be flattened; but their elasticity will make them instantaneously resume their former shape.

CAROLINE.

But when two ivory balls strike against each other, as they constantly do on a billiard table, no mark or impression is made by the stroke.

MRS. B.

I beg your pardon; but you cannot perceive any mark, because their elasticity instantly destroys all trace of it.

Soft bodies, which easily retain impressions, such as clay, wax, tallow, butter, &c. have very little elasticity; but of all descriptions of bodies liquids are the least elastic.

EMILY.

If sealing-wax were elastic, instead of retaining

the impression of a seal, it would resume a
smooth surface as soon as the weight of the seal
was removed.    But pray what is it that produces
the elasticity of bodies ?

There is great diversity of opinion upon that
point, and I cannot pretend to decide which approaches nearest to the truth.    Elasticity implies
susceptibility of compression, and the susceptibility
of compression depends upon the porosity of bodies,
for were there no pores or spaces between the particles of matter of which a body is composed, it
could not be compressed.

That is to say, that if the particles of bodies were
as close together as possible, they could not be
squeezed closer.

Bodies then, whose particles are most distant
from each other, must be most susceptible of compression, and consequently most elastic; and
this you say is the case with air, which is perhaps
the least dense of all bodies ?

You will not in general find this rule hold good,
for liquids have scarcely any elasticity, whilst hard

15

bodies are eminent for this property, though the latter are certainly of much greater density than the former; elasticity implies, therefore, not only a susceptibility of compression, but depends upon the power of resuming its former state after compression.

CAROLINE.

But surely there can be no pores in ivory and metals, Mrs. B.; how then can they be susceptible of compression?

MRS. B.

The pores of such bodies are invisible to the naked eye, but you must not thence conclude that they have none; it is, on the contrary, well ascertained that gold, one of the most dense of all bodies, is extremely porous, and that these pores are sufficiently large to admit water when strongly compressed to pass through them. This was shown by a celebrated experiment made many years ago at Florence.

EMILY.

If water can pass through gold, there must certainly be pores or interstices which afford it a passage; and if gold is so porous, what must other bodies be, which are so much less dense than gold!

MRS. B.

The chief difference in this respect is, I believe, that the pores in some bodies are larger than in others;

in cork, sponge, and bread, they form considerable
cavities; in wood and stone, when not polished,
they are generally perceptible to the naked eye;
whilst in ivory, metals, and all varnished and polished
bodies, they cannot be discerned.  To give you an
idea of the extreme porosity of bodies, Sir Isaac
Newton conjectured that if the earth were so compressed as to be absolutely without pores, its dimensions might possibly not be more than a cubic inch.

### CAROLINE.

What an idea !  Were we not indebted to Sir
Isaac Newton for the theory of attraction, I should
be tempted to laugh at him for such a supposition.
What insignificant little creatures we should be !

### MRS. B.

If our consequence arose from the size of our
bodies we should indeed be but pigmies, but remember that the mind of Newton was not circumscribed
by the dimensions of its envelope.

### EMILY.

It is, however, fortunate that heat keeps the pores
of matter open and distended, and prevents the
attraction of cohesion from squeezing us into a
nutshell.

### MRS. B.

Let us now return to the subject of re-action, on
which we have some further observations to make.

It is re-action, being contrary to action which produces *reflected motion.* If you throw a ball against the wall, it rebounds; this return of the ball is owing to the re-action of the wall against which it struck, and is called *reflected motion.*

EMILY.

And I now understand why balls filled with air rebound better than those stuffed with bran or wool; air being most susceptible of compression and most elastic, the re-action is more complete.

CAROLINE.

I have observed that when I throw a ball straight against the wall, it returns straight to my hand ; but if I throw it obliquely upwards, it rebounds still higher, and I catch it when it falls.

MRS. B.

You should not say straight, but perpendicularly against the wall; for straight is a general term for lines in all directions which are neither curved nor bent, and is therefore equally applicable to oblique or perpendicular lines.

CAROLINE.

I thought that perpendicularly meant either directly upwards or downwards?

E

74 ON THE LAWS OF MOTION.

In those directions lines are perpendicular to the
earth. A perpendicular line has always a reference
to something towards which it is perpendicular;
that is to say, that it inclines neither to the one
side or the other, but makes an equal angle on
every side. Do you understand what an angle is?

CAROLINE.

Yes, I believe so: it is two lines meeting in a
point.

MRS. B.

Well then, let the line A B (plate II, fig. 1.)
represent the floor of the room, and the line C D
that in which you throw a ball against it; the line
C D, you will observe, forms two angles with the
line A B, and those two angles are equal.

EMILY.

How can the angles be equal, while the lines
which compose them are of unequal length?

MRS. B.

An angle is not measured by the length of the
lines, but by their opening.

EMILY.

Yet the longer the lines are, the greater is the
opening between them.

PLATE II.

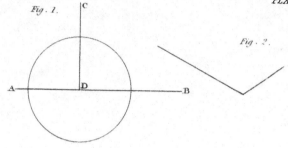

Fig. 1.

Fig. 2.

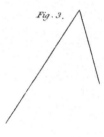

Fig. 3.

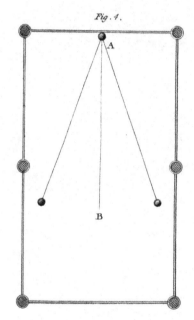

Fig. 4.

Fig. 5.

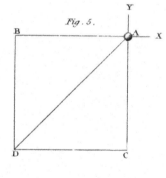

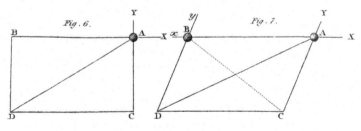

Fig. 6.

Fig. 7.

Published by Longman & Co. June 1st 1840.

Lowry Sc.

MRS. B.

Take a pair of compasses and draw a circle over these angles, making the angular point the centre.

EMILY.

To what extent must I open the compasses?

MRS. B.

You may draw the circle what size you please, provided that it cuts the lines of the angles we are to measure.    All circles, of whatever dimensions, are supposed to be divided into 360 equal parts, called degrees; the opening of an angle, being therefore a portion of a circle, must contain a certain number of degrees; the larger the angle, the greater the number of degrees, and two angles are said to be equal when they contain an equal number of degrees.

EMILY.

Now I understand it.   As the dimensions of an angle depend upon the number of degrees contained between its lines, it is the opening, and not the length of its lines, which determines the size of the angle.

MRS. B.

Very well: now that you have a clear idea of the dimensions of angles, can you tell me how many degrees are contained in the two angles formed by

one line falling perpendicularly on another, as in
the figure I have just drawn?

### EMILY.

You must allow me to put one foot of the
compasses at the point of the angles, and draw a
circle round them, and then I think I shall be able
to answer your question: the two angles are together
just equal to half a circle, they contain therefore
90 degrees each; 90 degrees being a quarter of 360.

### MRS. B.

An angle of 90 degrees is called a right angle,
and when one line is perpendicular to another, it
forms, you see, (fig. 1.) a right angle on either
side.    Angles containing more than 90 degrees
are called obtuse angles (fig. 2.); and those containing less than 90 degrees are called acute angles,
(fig. 3.)

### CAROLINE.

The angles of this square table are right angles,
but those of the octagon table are obtuse angles;
and the angles of sharp-pointed instruments are
acute angles.

### MRS. B.

Very well.  To return now to your observation,
that if a ball is thrown obliquely against the wall it
will not rebound in the same direction;  tell me,
have you ever played at billiards?

CAROLINE.

Yes, frequently; and I have observed that when I push the ball perpendicularly against the cushion, it returns in the same direction; but when I send it obliquely to the cushion, it rebounds obliquely, but on the opposite side; the ball in this latter case describes an angle, the point of which is at the cushion. I have observed too, that the more obliquely the ball is struck against the cushion, the more obliquely it rebounds on the opposite side, so that a billiard player can calculate with great accuracy in what direction it will return.

MRS. B.

Very well. This figure (fig. 4. plate II.) represents a billiard table; now if you draw a line A B from the point where the ball A strikes perpendicular to the cushion; you will find that it will divide the angle which the ball describes into two parts, or two angles; the one will show the obliquity of the direction of the ball in its passage towards the cushion, the other its obliquity in its passage back from the cushion. The first is called *the angle of incidence*, the other *the angle of reflection*, and these angles are always equal.

CAROLINE.

This then is the reason why, when I throw a ball obliquely against the wall, it rebounds in an opposite oblique direction, forming equal angles of incidence and of reflection.

MRS. B.

Certainly; and you will find that the more obliquely you throw the ball, the more obliquely it will rebound.

We must now conclude; but I shall have some further observations to make upon the laws of motion, at our next meeting.

# CONVERSATION IV.

## ON COMPOUND MOTION.

COMPOUND MOTION, THE RESULT OF TWO OPPOSITE
FORCES. — OF CIRCULAR MOTION, THE RESULT OF
TWO FORCES, ONE OF WHICH CONFINES THE BODY
TO A FIXED POINT. — CENTRE OF MOTION, THE
POINT AT REST WHILE THE OTHER PARTS OF THE
BODY MOVE ROUND IT. — CENTRE OF MAGNITUDE,
THE MIDDLE OF A BODY. — CENTRIPETAL FORCE,
THAT WHICH CONFINES A BODY TO A FIXED CEN-
TRAL POINT. — CENTRIFUGAL FORCE, THAT WHICH
IMPELS A BODY TO FLY FROM THE CENTRE. — FALL
OF BODIES IN A PARABOLA. — CENTRE OF GRAVITY,
THE CENTRE OF WEIGHT, OR POINT ABOUT WHICH
THE PARTS BALANCE EACH OTHER.

MRS. B.

I MUST now explain to you the nature of compound motion. Let us suppose a body to be struck
by two equal forces in opposite directions, how will
it move?

E 4

If the directions of the forces are in exact opposition to each other, I suppose the body would not move at all.

You are perfectly right; but if the forces, instead of acting on the body in opposition, strike it in two directions inclined to each other, at an angle of ninety degrees, if the ball A (fig. 5. plate II.) be struck by equal forces at X and at Y, will it not move?

The force X would send it towards B, and the force Y towards C; and since these forces are equal, I do not know how the body can obey one impulse rather than the other, and yet I think the ball would move, because as the two forces do not act in direct opposition, they cannot entirely destroy the effect of each other.

Very true; the ball will therefore follow the direction of neither of the forces, but will move in a line between them, and will reach D in the same space of time, that the force X would have sent it to B, and the force Y would have sent it to C. Now if you draw two lines from D, to join B and C, you will form a square, and the oblique line which the body describes is called the diagonal of the square.

That is very clear, but supposing the two forces to be unequal, that the force X, for instance, be twice as great as the force Y?

MRS. B.

Then the force X would drive the ball twice as far as the force Y, consequently you must draw the line A B (fig. 6.), twice as long as the line A C, the body will in this case move to D; and if you draw lines from that point to B and C, you will find that the ball has moved in the diagonal of a rectangle.

EMILY.

Allow me to put another case? Suppose the two forces are unequal, but do not act on the ball in the direction of a right angle, but in that of an acute angle, what will result?

MRS. B.

Prolong the lines in the directions of the two forces, and you will soon discover which way the ball will be impelled; it will move from A to D, in the diagonal of a parallelogram. (fig. 7.) Forces acting in the direction of lines forming an obtuse angle, will also produce motion in the diagonal of a parallelogram. For instance, if the body set out from B, instead of A, and was impelled by the forces X and Y, it would move in the dotted diagonal B C.

We may now proceed to circular motion: this is the result of two forces on a body, by one of which it is projected forward in a right line, whilst by the other it is confined to a fixed point.   For instance; when I whirl this ball, which is fastened to my hand with a string, the ball moves in a circular direction; because it is acted on by two forces, that which I give it which represents the force of projection, and that of the string which confines it to my hand.   If during its motion you were suddenly to cut the string, the ball would fly off in a straight line; being released from confinement to the fixed point, it would be acted on but by one force, and motion produced by one force, you know, is always in a right line.

### CAROLINE.

This is a little more difficult to comprehend than compound motion in straight lines.

### MRS. B.

You have seen a mop trundled, and have observed, that the threads which compose the head of the mop fly from the centre; but being confined to it at one end, they cannot part from it; whilst the water they contain, being unconfined, is thrown off in straight lines.

EMILY.

In the same way, the flyers of a windmill, when put in motion by the wind, would be driven straight forwards in a right line, were they not confined to a fixed point, round which they are compelled to move.

MRS. B.

Very well. And observe, that the point to which the motion of a small body, such as the ball with the string, which may be considered as revolving in one plane, is confined, becomes the centre of its motion. But when the bodies are not of a size or shape to allow of our considering every part of them as moving in the same plane, they in reality revolve round a line, which line is called the *axis of motion.* In a top, for instance, when spinning on its point, the axis is the line which passes through the middle of it, perpendicularly to the floor.

CAROLINE.

The axle of the flyers of the windmill, is then the axis of its motion; but is the centre of motion always in the middle of a body?

MRS. B.

No, not always. The middle point of a body, is called its centre of magnitude, or position, that is, the centre of its mass or bulk. Bodies have also another centre, called the centre of gravity, which

E 6

I shall explain to you; but at present, we must confine ourselves to the axis of motion. This line you must observe remains at rest, whilst all the other parts of the body move around it; when you spin a top the axis is stationary whilst every other part is in motion round it.

CAROLINE.

But a top generally has a motion forwards, besides its spinning motion; and then no point within it can be at rest?

MRS. B.

What I say of the axis of motion, relates only to circular motion; that is to say, to motion round a line, and not to that which a body may have at the same time in any other direction. There is one circumstance in circular motion, which you must carefully attend to; which is, that the further any part of a body is from the axis of motion, the greater is its velocity; as you approach that line, the velocity of the parts gradually diminish till you reach the axis of motion, which is perfectly at rest.

CAROLINE.

But, if every part of the same body did not move with the same velocity, that part which moved quickest, must be separated from the rest of the body, and leave it behind?

PLATE III

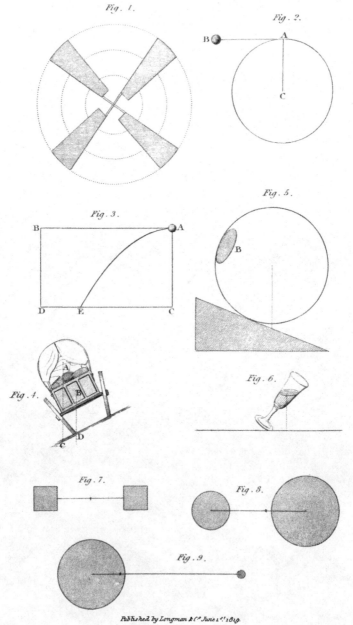

Fig. 1.

Fig. 2.

Fig. 3.

Fig. 4.

Fig. 5.

Fig. 6.

Fig. 7.

Fig. 8.

Fig. 9.

Published by Longman & Co. June 1.st 1819.

Lowry Sc.

MRS. B.

You perplex yourself by confounding the idea of circular motion, with that of motion in a right line; you must think only of the motion of a body round a fixed line, and you will find, that if the parts farthest from the centre had not the greatest velocity, those parts would not be able to keep up with the rest of the body, and would be left behind. Do not the extremities of the vanes of a windmill move over a much greater space, than the parts nearest the axis of motion? (plate III. fig. 1.) the three dotted circles describe the paths in which three different parts of the vanes move, and though the circles are of different dimensions, the vanes describe each of them in the same space of time.

CAROLINE.

Certainly they do; and I now only wonder, that we neither of us ever made the observation before: and the same effect must take place in a solid body, like the top in spinning; the most bulging part of the surface must move with the greatest rapidity.

MRS. B.

The force which confines a body to a centre, round which it moves is called the *centripetal* force; and that force, which impels a body to fly from the centre, is called the *centrifugal* force; in circular motion, these two forces constantly balance each other; otherwise the revolving body would either

approach the centre or recede from it, according as
the one or the other prevailed.

### CAROLINE.

When I see any body moving in a circle, I shall
remember, that it is acted on by two forces.

### MRS. B.

Motion, either in a circle, an ellipsis, or any other
curve-line, must be the result of the action of two
forces; for you know, that the impulse of one
single force, always produces motion in a right line.

### EMILY.

And if any cause should destroy the centripetal
force, the centrifugal force would alone impel the
body, and it would I suppose fly off in a straight
line from the centre to which it had been confined.

### MRS. B.

It would not fly off in a right line from the centre; but in a right line in the direction in which it
was moving, at the instant of its release; if a stone,
whirled round in a sling, gets loose at the point A,
(plate III. fig. 2.) it flies off in the direction A B;
this line is called a *tangent*, it touches the circumference of the circle, and forms a right angle with
a line drawn from that point of the circumference
to the centre of the circle, C.

You say, that motion in a curve-line, is owing to two forces acting upon a body; but when I throw this ball in an horizontal direction, it describes a curve line in falling; and yet it is only acted upon by the force of projection; there is no centripetal force to confine it, or produce compound motion.

MRS. B.

A ball thus thrown, is acted upon by no less than three forces; the force of projection, which you communicated to it; the resistance of the air through which it passes, which diminishes its velocity, without changing its direction; and the force of gravity, which finally brings it to the ground. The power of gravity, and the resistance of the air, being always greater than any force of projection we can give a body, the latter is gradually overcome, and the body brought to the ground; but the stronger the projectile force, the longer will these powers be in subduing it, and the further the body will go before it falls.

CAROLINE.

A shot fired from a cannon, for instance, will go much further, than a stone projected by the hand.

MRS. B.

Bodies thus projected, you observed, described a

curve-line in their descent; can you account for
that?

No; I do not understand, why it should not fall
in the diagonal of a square.

You must consider that the force of projection is
strongest when the ball is first thrown; this force,
as it proceeds, being weakened by the continued
resistance of the air, the stone, therefore, begins
by moving in an horizontal direction; but as the
stronger powers prevail, the direction of the ball
will gradually change from an horizontal, to a perpendicular line.    Projection alone, would drive the
ball A, to B, (fig. 3.) gravity would bring it to C;
therefore, when acted on in different directions, by
these two forces, it moves between, gradually in-
clining more and more to the force of gravity, in
proportion as this accumulates; instead therefore
of reaching the ground at D, as you supposed it
would, it falls somewhere about E.

It is precisely so; look, Emily, as I throw this
ball directly upwards, how the resistance of the air
and gravity conquers projection.    Now I will throw
it upwards obliquely; see, the force of projection

13

enables it, for an instant, to act in opposition to that of gravity; but it is soon brought down again.

### MRS. B.

The curve-line which the ball has described, is called in geometry a *parabola;* but when the ball is thrown perpendicularly upwards, it will descend perpendicularly; because the force of projection, and that of gravity, are in the same line of direction.

We have noticed the centres of magnitude, and of motion; but I have not yet explained to you, what is meant by the centre of gravity; it is that point in a body, about which all the parts exactly balance each other; if therefore that point is supported, the body will not fall. Do you understand this?

### EMILY.

I think so, if the parts round about this point have an equal tendency to fall, they will be in equilibrium, and as long as this point is supported, the body cannot fall.

### MRS. B.

Caroline, what would be the effect, were any other point of the body alone supported?

### CAROLINE.

The surrounding parts no longer balancing each

other, the body, I suppose, would fall on the side
at which the parts are heaviest.

Infallibly; whenever the centre of gravity is unsupported, the body must fall. This sometimes
happens with an overloaded waggon winding up a
steep hill, one side of the road being more elevated
than the other; let us suppose it to slope as is
described in this figure, (plate III. fig. 4.,) we will
say, that the centre of gravity of this loaded waggon is at the point A. Now your eye will tell
you, that a waggon thus situated, will overset; and
the reason is, that the centre of gravity A, is not
supported; for if you draw a perpendicular line
from it to the ground at C, it does not fall under
the waggon within the wheels, and is therefore not
supported by them.

I understand that perfectly; but what is the
meaning of the other point B?

Let us, in imagination take off the upper part of
the load; the centre of gravity will then change its
situation, and descend to B, as that will now be
the point about which the parts of the less heavily

laden waggon will balance each other. Will the waggon now be upset?

CAROLINE.

No, because a perpendicular line from that point falls within the wheels at D, and is supported by them; and when the centre of gravity is supported, the body will not fall.

EMILY.

Yet I should not much like to pass a waggon, in that situation; for, as you see, the point D is but just within the left wheel; if the right wheel was merely raised, by passing over a stone, the point D would be thrown on the outside of the left wheel, and the waggon would upset.

CAROLINE.

A waggon, or any carriage whatever, will then be most firmly supported, when the centre of gravity falls exactly between the wheels; and that is the case in a level road.

Pray, whereabouts is the centre of gravity of the human body?

MRS. B.

Between the hips; and as long as we stand upright, this point is supported by the feet; if you lean on one side, you will find that you no longer stand firm. A rope-dancer performs all his feats of agility, by dexterously supporting his centre of gravity; whenever he finds, that he is in danger of losing his balance, he shifts the heavy pole, which he holds in his hands, in order to throw the weight towards the side that is deficient; and thus by changing the situation of the centre of gravity, he restores his equilibrium.

CAROLINE.

When a stick is poised on the tip of the finger, is it not by supporting its centre of gravity?

MRS. B.

Yes; and it is because the centre of gravity is not supported, that spherical bodies roll down a slope. A sphere, being perfectly round, can touch the slope but by a single point, and that point cannot be perpendicularly under the centre of gravity, and therefore cannot be supported, as you will perceive by examining this figure. (Fig. 5. plate III.)

EMILY.

So it appears; yet I have seen a cylinder of wood roll up a slope; how is that contrived?

MRS. B.

It is done by plugging one side of the cylinder with lead, as at B, (fig. 5. plate III.) the body being no longer of an uniform density, the centre of gravity is removed from the middle of the body to some point in the lead, as that substance is much heavier than wood; now you may observe that in order that the cylinder may roll down the plane, as it is here situated, the centre of gravity must rise, which is impossible; the centre of gravity must always descend in moving, and will descend by the nearest and readiest means, which will be by forcing the cylinder up the slope, until the centre of gravity is supported, and then it stops.

CAROLINE.

The centre of gravity, therefore, is not always in the middle of a body?

MRS. B.

No, that point we have called the centre of magnitude; when the body is of an uniform density, the centre of gravity is in the same point; but when one part of the body is composed of heavier materials than another part, the centre of gravity being the centre of the weight of the body can no longer correspond with the centre of magnitude. Thus you see the centre of gravity of this cylinder plugged with lead, cannot be in the same spot as the centre of magnitude.

EMILY.

Bodies, therefore, consisting but of one kind of

substance, as wood, stone, or lead, and whose densities are consequently uniform, must stand more firmly, and be more difficult to overset, than bodies composed of a variety of substances, of different densities, which may throw the centre of gravity on one side.

MRS. B.

Yes; but there is another circumstance which more materially affects the firmness of their position, and that is their form. Bodies that have a narrow base are easily upset, for if they are the least inclined, their centre is no longer supported, as you may perceive in fig. 6.

CAROLINE.

I have often observed with what difficulty a person carries a single pail of water; it is owing, I suppose, to the centre of gravity being thrown on one side, and the opposite arm is stretched out to endeavour to bring it back to its original situation; but a pail hanging on each arm is carried without difficulty, because they balance each other, and the centre of gravity remains supported by the feet.

MRS. B.

Very well; I have but one more remark to make on the centre of gravity, which is, that when two bodies are fastened together, by a line, string, chain, or any power whatever, they are to be considered as forming but one body; if the two bodies be of equal weight, the centre of gravity will be in the middle of the line which unites them, (fig. 7.) but if one be heavier than the other, the centre of gravity will be proportionally nearer the heavy body than the light one. (fig. 8.) If you were to carry a rod or pole with an equal weight fastened at each end of it, you would hold it in the middle of the rod, in order that the weights should balance each other; whilst if it had unequal weights at each end, you would hold it nearest the greater weight, to make them balance each other.

EMILY.

And in both cases we should support the centre of gravity; and if one weight be very considerably larger than the other, the centre of gravity will be thrown out of the rod into the heaviest weight. (fig. 9.)

MRS. B.

Undoubtedly.

# CONVERSATION V.

## ON THE MECHANICAL POWERS.

OF THE POWER OF MACHINES. — OF THE LEVER
IN GENERAL. — OF THE LEVER OF THE FIRST
KIND, HAVING THE FULCRUM BETWEEN THE
POWER AND THE WEIGHT. — OF THE LEVER OF
THE SECOND KIND, HAVING THE WEIGHT BE-
TWEEN THE POWER AND THE FULCRUM. — OF
THE LEVER OF THE THIRD KIND, HAVING THE
POWER BETWEEN THE FULCRUM AND THE
WEIGHT.

MRS. B.

WE may now proceed to examine the mechanical
powers; they are six in number, one or more of
which enters into the composition of every machine.
The *lever*, the *pulley*, the *wheel* and *axle*, the *inclined plane*, the *wedge*, and the *screw*.

In order to understand the power of a machine,
there are four things to be considered. 1st. The

power that acts: this consists in the effort of men or horses, of weights, springs, steam, &c.

2dly. The resistance which is to be overcome by the power; this is generally a weight to be moved. The power must always be superior to the resistance, otherwise the machine could not be put in motion.

### CAROLINE.

If for instance the resistance of a carriage was greater than the strength of the horses employed to draw it, they would not be able to make it move.

### MRS. B.

3dly. We are to consider the centre of motion, or as it is termed in mechanics, the *fulcrum ;* this you may recollect is the point about which all the parts of the body move; and lastly, the respective velocities of the power, and of the resistance.

### EMILY.

That must depend upon their respective distances from the axis of motion; as we observed in the motion of the vanes of the windmill.

### MRS. B.

We shall now examine the power of the lever. The lever is an inflexible rod or beam of any kind, that is to say, one which will not bend in any direction. For instance, the steel rod to

which these scales are suspended is a lever, and
the point in which it is supported the fulcrum, or
centre of motion; now, can you tell me why the
two scales are in equilibrium?

CAROLINE.

Being both empty, and of the same weight, they
balance each other.

EMILY.

Or, more correctly speaking, because the centre
of gravity common to both is supported.

MRS. B.

Very well; and which is the centre of gravity of
this pair of scales? (fig. 1. plate III.)

EMILY.

You have told us that when two bodies of equal
weight were fastened together, the centre of gravity
was in the middle of the line that connected them;
the centre of gravity of the scales must therefore
be in the fulcrum F of the lever which unites the
two scales; and corresponds with the centre of
motion.

CAROLINE.

But if the scales contained different weights, the
centre of gravity would no longer be in the fulcrum
of the lever, but removed towards that scale which

11

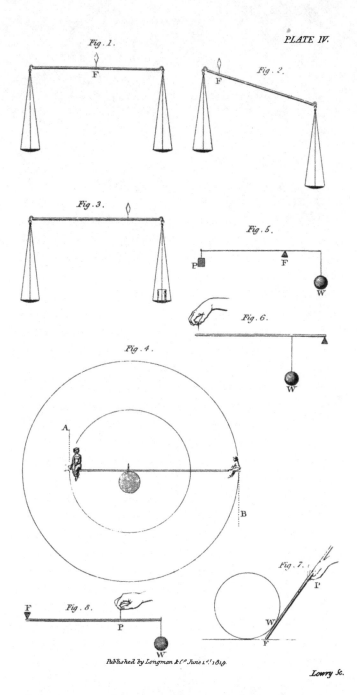

*Fig. 1.*

PLATE IV.

*Fig. 2.*

*Fig. 3.*

*Fig. 5.*

*Fig. 6.*

*Fig. 4.*

*Fig. 7.*

*Fig. 8.*

Published by Longman & C.º June 1.ˢᵗ 1819.

Lowry Sc.

contained the heaviest weight; and since that point would no longer be supported, the heavy scale would descend and out-weigh the other.

MRS. B.

True; but tell me, can you imagine any mode by which bodies of different weights can be made to balance each other, either in a pair of scales, or simply suspended to the extremities of the lever? for the scales are not an essential part of the machine, they have no mechanical power, and are used merely for the convenience of containing the substance to be weighed.

CAROLINE.

What! make a light body balance a heavy one? I cannot conceive that possible.

MRS. B.

The fulcrum of this pair of scales (fig. 2.) is moveable, you see; I can take it off the prop, and fasten it on again in another part; this part is now become the fulcrum, but it is no longer in the centre of the lever.

CAROLINE.

And the scales are no longer true; for that which hangs on the longest side of the lever descends.

F 2

MRS. B.

The two parts of the lever divided by the fulcrum are called its arms, you should therefore say the longest arm, not the longest side of the lever. These arms are likewise frequently distinguished by the appellations of the acting and the resisting part of the lever.

Your observation is true that the balance is now destroyed; but it will answer the purpose of enabling you to comprehend the power of a lever when the fulcrum is not in the centre.

EMILY.

This would be an excellent contrivance for those who cheat in the weight of their goods; by making the fulcrum a little on one side, and placing the goods in the scale which is suspended to the longest arm of the lever, they would appear to weigh more than they do in reality.

MRS. B.

You do not consider how easily the fraud would be detected; for on the scales being emptied they would not hang in equilibrium.

EMILY.

True; I did not think of that circumstance. But I do not understand why the longest arm of the lever should not be in equilibrium with the other?

CAROLINE.

It is because it is heavier than the shortest arm; the centre of gravity, therefore, is no longer supported.

MRS. B.

You are right; the fulcrum is no longer in the centre of gravity; but if we can contrive to make the fulcrum in its present situation become the centre of gravity, the scales will again balance each other; for you recollect that the centre of gravity is that point about which every part of the body is in equilibrium.

EMILY.

It has just occurred to me how this may be accomplished; put a great weight into the scale suspended to the shortest arm of the lever, and a smaller one into that suspended to the longest arm. Yes, I have discovered it — look, Mrs. B., the scale on the shortest arm will carry 2lbs., and that on the longest arm only one, to restore the balance. (fig. 3.)

MRS. B.

You see, therefore, that it is not so impracticable as you imagined to make a heavy body balance a light one; and this is in fact the means by which you thought an imposition in the weight of goods might be effected, as a weight of ten or twelve ounces might thus be made to balance a pound of goods. Let us now take off the scales that we may consider the lever simply; and in this state you see that the fulcrum is no longer the centre of gravity; but it is, and must ever be, the centre of motion, as it is the only point which remains at rest, while the other parts move about it.

### CAROLINE.

It now resembles the two opposite vanes of a windmill, and the fulcrum the point round which they move.

### MRS. B.

In describing the motion of those vanes, you may recollect our observing that the farther a body is from the axis of motion the greater is its velocity.

### CAROLINE.

That I remember and understood perfectly.

### MRS. B.

You comprehend then, that the extremity of the longest arm of a lever must move with greater velocity than that of the shortest arm?

### EMILY.

No doubt, because it is farthest from the centre of motion. And pray, Mrs. B., when my brothers play at *see-saw,* is not the plank on which they ride a kind of lever?

### MRS. B,

Certainly; the log of wood which supports it from the ground is the fulcrum, and those who ride represent the power and the resistance at each end of the lever.   And have you not observed that when those who ride are of equal weight, the plank must be supported in the middle to make the two arms equal; whilst if the persons differ in weight, the plank must be drawn a little further over the prop, to make the arms unequal, and the lightest person who represents the resistance, must be placed at the extremity of the longest arm.

### CAROLINE.

That is always the case when I ride on a plank with my youngest brother; I have observed also that the lightest person has the best ride, as he moves both further and quicker; and I now understand that it is because he is more distant from the centre of motion.

### MRS. B.

The greater velocity with which your little brother moves, renders his momentum equal to yours.

## CAROLINE.

Yes; I have the most gravity, he the greatest velocity; so that upon the whole our momentums are equal.—But you said, Mrs. B., that the power should be greater than the resistance to put the machine in motion; how then can the plank move if the momentums of the persons who ride are equal.

## MRS. B.

Because each person at his descent touches the ground with his feet; the reaction of which gives him an impulse which increases his velocity; this spring is requisite to destroy the equilibrium of the power and the resistance, otherwise, the plank would not move.   Did you ever observe that a lever describes the arc of a circle in its motion?

## EMILY.

No; it appears to me to rise and descend perpendicularly; at least I always thought so.

## MRS. B.

I believe I must make a sketch of you and your brother riding on a plank in order to convince you of your error. (fig. 4. pl. IV.)  You may now observe that a lever can move only round the fulcrum, since that is the centre of motion; it would be impossible for you to rise perpendicularly to the point A, or for your brother to descend in a straight line to the point B; you must in rising and he in descending describe arcs of your respective circles. This drawing shews you also how much superior his velocity must be to yours; for if you could swing quite round, you would each complete your respective circles in the same time.

### CAROLINE.

My brother's circle being much the largest, he must undoubtedly move the quickest.

### MRS. B.

Now tell me, do you think that your brother could raise you as easily without the aid of a lever?

### CAROLINE.

Oh no, he could not lift me off the ground.

### MRS. B.

Then I think you require no further proof of the power of a lever, since you see what it enables your brother to perform.

### CAROLINE.

I now understand what you meant by saying, that in mechanics, motion was opposed to matter, for it is my brother's velocity which overcomes my weight.

F 5

You may easily imagine, what enormous weights may be raised by levers of this description, for the longer the acting part of the lever in comparison to the resisting part, the greater is the effect produced by it; because the greater is the velocity of the power compared to that of the weight.

There are three different kinds of levers; in the first the fulcrum is between the power and the weight.

CAROLINE.

This kind then comprehends the several levers you have described.

MRS. B.

Yes, when in levers of the first kind, the fulcrum is equally between the power and the weight, as in the balance, the power must be greater than the weight, in order to move it; for nothing can in this case be gained by velocity; the two arms of the lever being equal, the velocity of their extremities must be so likewise. The balance is therefore of no assistance as a mechanical power, but it is extremely useful to estimate the respective weights of bodies.

But when (fig. 5.) the fulcrum F of a lever is not equally distant from the power and the weight, and that the power P acts at the extremity of the longest arm, it may be less than the weight W,

its deficiency being compensated by its superior velocity; as we observed in the *see-saw*.

### EMILY.

Then when we want to lift a great weight, we
must fasten it to the shortest arm of a lever, and
apply our strength to the longest arm?

### MRS. B.

If the case will admit of your putting the end
of the lever under the weight, no fastening will be
required; as you will perceive by stirring the fire.

### EMILY.

Oh yes! the poker is a lever of the first kind,
the point where it rests against the bars of the grate
whilst I am stirring the fire, is the fulcrum; the
short arm or resisting part of the lever, is employed
in lifting the weight, which is the coals, and my
hand is the power applied to the longest arm, or
acting part of the lever.

### MRS. B,

Let me hear, Caroline, whether you can equally
well explain this instrument, which is composed of
two levers, united in one common fulcrum.

### CAROLINE.

A pair of scissars!

F 6

You are surprised, but if you examine their construction, you will discover that it is the power of the lever that assists us in cutting with scissars.

CAROLINE.

Yes; I now perceive that the point at which the two levers are screwed together, is the fulcrum; the handles, to which the power of the fingers is applied, are the extremities of the acting part of the levers, and the cutting part of the scissars, are the resisting parts of the levers: therefore, the longer the handles and the shorter the points of the scissars, the more easily you cut with them.

EMILY.

That I have often observed, for when I cut pasteboard or any hard substance, I always make use of that part of the scissars nearest the screw or rivet, and I now understand why it increases the power of cutting; but I confess that I never should have discovered scissars to have been double levers; and pray are not snuffers levers of a similar description?

MRS. B.

Yes, and most kinds of pincers; the great power of which consists in the resisting part of the lever being very short in comparison of the acting part.

CAROLINE.

And of what nature are the two other kinds
of levers?

MRS. B.

In levers of the second kind, the weight, instead
of being at one end, is situated between the power
and the fulcrum, (fig.6.)

CAROLINE.

The weight and the fulcrum have here changed
places; and what advantage is gained by this kind
of lever?

MRS. B.

In moving it, the velocity of the power must
necessarily be greater than that of the weight, as it
is more distant from the centre of the motion.
Have you ever seen your brother move a snow-
ball by means of a strong stick, when it became too
heavy for him to move without assistance?

CAROLINE.

Oh yes; and this was a lever of the second order
(fig. 7.); the end of the stick, which he thrusts under
the ball, and which rests on the ground, becomes
the fulcrum; the ball is the weight to be moved,
and the power his hands applied to the other end
of the lever. In this instance there is an immense
difference in the length of the arms of the lever;
for the weight is almost close to the fulcrum.

And the advantage gained is proportional to this difference. Fishermen's boats are by levers of this description raised from the ground to be launched into the sea, by means of slippery pieces of board which are thrust under the keel. The most common example that we have of levers of the second kind is in the doors of our apartments.

The hinges represent the fulcrum, our hands the power applied to the other end of the lever; but where is the weight to be moved?

The door is the weight, and it consequently occupies the whole of the space between the power and the fulcrum. Nutcrackers are double levers of this kind: the hinge is the fulcrum, the nut the resistance, and the hands the power.

In levers of the third kind (fig. 8.), the fulcrum is again at one of the extremities, the weight or resistance at the other, and it is now the power which is applied between the fulcrum and the resistance.

The fulcrum, the weight, and the power, then, each in their turn, occupy some part of the middle of the lever between its extremities. But in this third kind of lever, the weight being farther from the centre of motion than the power, the difficulty of raising it seems increased rather than diminished.

MRS. B.

That is very true; a lever of this kind is therefore never used, unless absolutely necessary, as is the case in lifting up a ladder perpendicularly in order to place it against a wall; the man who raises it cannot place his hands on the upper part of the ladder, the power, therefore, is necessarily placed much nearer the fulcrum than the weight.

CAROLINE.

Yes, the hands are the power, the ground the fulcrum, and the upper part of the ladder the weight.

MRS. B.

Nature employs this kind of lever in the structure of the human frame. In lifting a weight with the hand, the lower part of the arm becomes a lever of the third kind; the elbow is the fulcrum, the muscles of the fleshy part of the arm the power; and as these are nearer to the elbow than the hand, it is necessary that their power should exceed the weight to be raised.

EMILY.

Is it not surprising that nature should have furnished us with such disadvantageous levers?

MRS. B.

The disadvantage, in respect to power, is more than counterbalanced by the convenience resulting from this structure of the arm; and it is no doubt that which is best adapted to enable it to perform its various functions.

We have dwelt so long on the lever, that we must reserve the examination of the other mechanical powers to our next interview.

PLATE V.

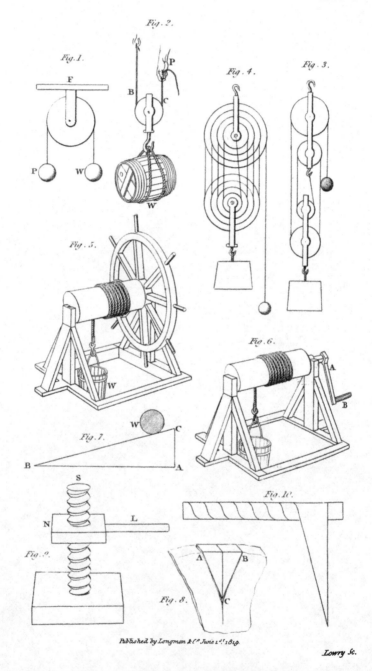

Fig. 1.
Fig. 2.
Fig. 4.
Fig. 3.
Fig. 5.
Fig. 6.
Fig. 7.
Fig. 9.
Fig. 8.
Fig. 10.

Published by Longman & Co. June 1st 1849.

Lowry Sc.

# CONVERSATION VI.

## ON THE MECHANICAL POWERS.

OF THE PULLEY. — OF THE WHEEL AND AXLE. —
OF THE INCLINED PLANE. — OF THE WEDGE. —
OF THE SCREW.

MRS. B.

THE pulley is the second mechanical power we
are to examine. You both, I suppose, have seen a
pulley?

CAROLINE.

Yes, frequently: it is a circular and flat piece of
wood or metal, with a string which runs in a groove
round it; by means of which, a weight may be
pulled up; thus pulleys are used for drawing up
curtains.

MRS. B.

Yes; but in that instance the pulleys are fixed,
and do not increase the power to raise the weights,
as you will perceive by this figure. (plate V. fig. 1.)
Observe that the fixed pulley is on the same princi-

ple as the lever of a pair of scales, in which the fulcrum F being in the centre of gravity, the power P
and the weight W, are equally distant from it, and
no advantage is gained.

EMILY.

Certainly; if P represents the power employed
to raise the weight W, the power must be greater
than the weight in order to move it. But of what
use then are pulleys in mechanics?

MRS. B.

The next figure represents a pulley which is not
fixed, (fig. 2.) and thus situated you will perceive
that it affords us mechanical assistance. In order
to raise the weight (W) one inch, P, the power,
must draw the strings B and C one inch each; the
whole string is therefore shortened two inches,
while the weight is raised only one.

EMILY.

That I understand: if P drew the string but one
inch, the weight would be raised only half an inch,
because it would shorten the strings B and C half
an inch each, and consequently the pulley, with the
weight attached to it, can be raised only half an
inch.

CAROLINE.

I am ashamed of my stupidity; but I confess that

I do not understand this; it appears to me that the weight would be raised as much as the string is shortened by the power.

### MRS. B.

I will endeavour to explain it more clearly. I fasten this string to a chair and draw it towards me; I have now shortened the string, by the act of drawing it, one yard.

### CAROLINE.

And the chair, as I supposed, has advanced one yard.

### MRS. B.

This exemplifies the nature of a single fixed pulley only. Now unfasten the string, and replace the chair where it stood before. In order to represent the moveable pulley, we must draw the chair forwards by putting the string round it; one end of the string may be fastened to the leg of the table, and I shall draw the chair by the other end of the string. I have again shortened the string one yard; how much has the chair advanced?

### CAROLINE.

I now understand it; the chair represents the weight to which the moveable pulley is attached; and it is very clear that the weight can be drawn only half the length you draw the string. I believe the circumstance that perplexed me was, that I did not observe the difference that results from the weight being attached to the pulley, instead of being fastened to the string, as is the case in the fixed pulley.

EMILY.

But I do not yet understand the advantage of pulleys; they seem to me to increase rather than diminish the difficulty of raising weights, since you must draw the string double the length that you raise the weight; whilst with a single pulley, or without any pulley, the weight is raised as much as the string is shortened.

MRS. B.

The advantage of a moveable pulley consists in dividing the difficulty; we must draw, it is true, twice the length of the string, but then only half the strength is required that would be necessary to raise the weight without the assistance of a moveable pulley.

EMILY.

So that the difficulty is overcome in the same manner as it would be, by dividing the weight into two equal parts, and raising them successively.

MRS. B.

Exactly. You must observe, that with a moveable pulley the velocity of the power is double that of the weight, since the power P (fig. 2.) moves two inches whilst the weight W moves one inch; therefore the power need not be more than half the weight to make their momentums equal.

### CAROLINE.

Pulleys act then on the same principle as the lever, the deficiency of strength of the power being compensated by its superior velocity.

### MRS. B.

You will find, that all mechanical power is founded on the same principle.

### EMILY.

But may it not be objected to pulleys, that a longer time is required to raise a weight by their aid than without it; for what you gain in power you lose in time?

### MRS. B.

That, my dear, is the fundamental law in mechanics: it is the case with the lever, as well as the pulley; and you will find it to be so with all the other mechanical powers.

### CAROLINE.

I do not see any advantage in the mechanical powers then, if what we gain by them one way is lost another.

Since we are not able to increase our natural strength, is not that science of wonderful utility, by means of which we may reduce the resistance or weight of any body to the level of our strength? This the mechanical powers enable us to accomplish, by dividing the resistance of a body into parts which we can successively overcome.   It is true, as you observe, that it requires a sacrifice of time to attain this end, but you must be sensible how very advantageously it is exchanged for power: the utmost exertion we can make adds but little to our natural strength, whilst we have a much more unlimited command of time.   You can now understand, that the greater the number of pulleys connected by a string, the more easily the weight is raised, as the difficulty is divided amongst the number of strings, or rather of parts into which the string is divided by the pulleys.   Several pulleys thus connected, form what is called a system, or tackle of pulleys. (fig. 3.)   You may have seen them suspended from cranes to raise goods into warehouses, and in ships to draw up the sails.

EMILY.

But since a fixed pulley affords us no mechanical aid, why is it ever used?

MRS. B.

Though it does not increase our power, it is

frequently useful for altering its direction. A single pulley enables us to draw *up* a curtain, by drawing *down* the string connected with it; and we should be much at a loss to accomplish this simple oper-ation without its assistance.

CAROLINE.

There would certainly be some difficulty in ascending to the head of the curtain, in order to draw it up. Indeed, I now recollect having seen workmen raise small weights by this means, which seemed to answer a very useful purpose.

MRS. B.

In shipping, both the advantages of an increase of power and a change of direction, by means of pulleys, are united; for the sails are raised up the masts by the sailors on deck, from the change of direction which the pulley effects, and the labour is facilitated by the mechanical power of a combination of pulleys,

EMILY.

But the pulleys on ship-board do not appear to me to be united in the manner you have shown us.

MRS. B.

They are, I believe, generally connected, as de-scribed in figure 4, both for nautical, and a variety

of other purposes; but in whatever manner pulleys
are connected by a single string, the mechanical
power is the same.

The third mechanical power is the wheel and
axle.   Let us suppose (plate VI. fig. 5.) the weight
W to be a bucket of water in a well, which we raise
by winding the rope, to which it is attached, round
the axle; if this be done without a wheel to turn
the axle, no mechanical assistance is received.   The
axle without a wheel is as impotent as a single fixed
pulley, or a lever, whose fulcrum is in the centre;
but add the wheel to the axle, and you will immediately find the bucket is raised with much less
difficulty.   The velocity of the circumference of the
wheel is as much greater than that of the axle, as
it is further from the centre of motion; for the
wheel describes a great circle in the same space
of time that the axle describes a small one;
therefore the power is increased in the same proportion as the circumference of the wheel is greater
than that of the axle.   If the velocity of the wheel
is twelve times greater than that of the axle, a
power nearly twelve times less than the weight of
the bucket would be able to raise it.

#### EMILY.

The axle acts the part of the shorter arm of the
lever, the wheel that of the longer arm.

### CAROLINE.

In raising water, there is commonly, I believe, instead of a wheel attached to the axle, only a crooked handle, which answers the purpose of winding the rope round the axle, and thus raising the bucket.

### MRS. B.

In this manner (fig. 6.): now if you observe the dotted circle which the handle describes in winding up the rope, you will perceive that the branch of the handle A, which is united to the axle, represents the spoke of a wheel, and answers the purpose of an entire wheel; the other branch B affords no mechanical aid, merely serving as a handle to turn the wheel.

Wheels are a very essential part of most machines: they are employed in various ways; but, when fixed to the axle, their mechanical power is always the same; that is, as the circumference of the wheel exceeds that of the axle, so much will the energy of the power be increased.

### CAROLINE.

Then the larger the wheel the greater must be its effect.

### MRS. B.

Certainly. If you have ever seen any considerable mills or manufactures, you must have ad-

G

mired the immense wheel, the revolution of which puts the whole of the machinery into motion; and though so great an effect is produced by it, a horse or two has sufficient power to turn it; sometimes a stream of water is used for that purpose, but of late years, a steam-engine has been found both the most powerful and the most convenient mode of turning the wheel.

CAROLINE.

Do not the vanes of a windmill represent a wheel, Mrs. B.?

MRS. B.

Yes; and in this instance we have the advantage of a gratuitous force, the wind, to turn the wheel. One of the great benefits resulting from the use of machinery is, that it gives us a sort of empire over the powers of nature, and enables us to make them perform the labour which would otherwise fall to the lot of man. When a current of wind, a stream of water, or the expansive force of steam, performs our task, we have only to superintend and regulate their operations.

The fourth mechanical power is the inclined plane; this is nothing more than a slope, or declivity, frequently used to facilitate the drawing up of weights. It is not difficult to understand, that a weight may much more easily be drawn up a slope than it can be raised the same height per-

pendicularly.  But in this, as well as the other
mechanical powers, the facility is purchased by
a loss of time (fig. 7.); for the weight, instead
of moving directly from A to C, must move
from B to C, and as the length of the plane is to
its height, so much is the resistance of the weight
diminished.

EMILY.

Yes; for the resistance, instead of being confined to the short line A C, is spread over the long
line B C.

MRS. B.

The wedge, which is the next mechanical power,
is composed of two inclined planes (fig. 8.): you
may have seen wood-cutters use it to cleave wood.
The resistance consists in the cohesive attraction
of the wood, or any other body which the wedge
is employed to separate; and the advantage gained
by this power is in the proportion of half its width
to its length; for while the wedge forces asunder
the coherent particles of the wood to A and B, it
penetrates downwards as far as C.

EMILY.

The wedge, then, is rather a compound than a
distinct mechanical power, since it is composed of
two inclined planes.

G 2

### MRS. B.

It is so.  All cutting instruments are constructed upon the principle of the inclined plane, or the wedge: those that have but one edge sloped, like the chisel, may be referred to the inclined plane; whilst the axe, the hatchet, and the knife (when used to split asunder) are used as wedges.

### CAROLINE.

But a knife cuts best when it is drawn across the substance it is to divide.  We use it thus in cutting meat, we do not chop it to pieces.

### MRS. B.

The reason of this is, that the edge of a knife is really a very fine saw, and therefore acts best when used like that instrument.

The screw, which is the last mechanical power, is more complicated than the others.  You will see by this figure, (fig. 9.) that it is composed of two parts, the screw and the nut.  The screw S is a cylinder, with a spiral protuberance coiled round it, called the thread; the nut N is perforated to contain the screw, and the inside of the nut has a spiral groove, made to fit the spiral thread of the screw.

### CAROLINE.

It is just like this little box, the lid of which

screws on the box as you have described; but what
is this handle which projects from the nut?

MRS. B.

It is a lever, which is attached to the nut, without which the screw is never used as a mechanical
power; the nut with a lever L attached to it, is commonly called a winch.   The power of the screw,
complicated as it appears, is referable to one of the
most simple of the mechanical powers; which of
them do you think it is?

CAROLINE.

In appearance, it most resembles the wheel and
axle.

MRS. B.

The lever, it is true, has the effect of a wheel, as
it is the means by which you wind the nut round;
but the lever is not considered as composing a part
of the screw, though it is true, that it is necessarily
attached to it.   But observe, that the lever, considered as a wheel, is not fastened to the axle or
screw, but moves round it, and in so doing, the nut
either rises or descends, according to the way in which
you turn it.

EMILY.

The spiral thread of the screw resembles, I
think, an inclined plane: it is a sort of slope, by
means of which the nut ascends more easily than

G 3

it would do if raised perpendicularly; and it serves
to support it when at rest.

Very well: if you cut a slip of paper in the form
of an inclined plane, and wind it round your pencil, which will represent the cylinder, you will find
that it makes a spiral line, corresponding to the
spiral protuberance of the screw. (Fig. 10.)

Very true; the nut then ascends an inclined
plane, but ascends it in a spiral, instead of a straight
line: the closer the thread of the screw, the more
easy the ascent; it is like having shallow, instead
of steep steps to ascend.

Yes; excepting that the nut takes no steps, it gradually winds up or down; then observe, that the closer
the threads of the screw, the greater the number
of revolutions the winch must make; so that we
return to the old principle, — what is saved in
power is lost in time.

Cannot the power of the screw be increased also,
by lengthening the lever attached to the nut?

MRS. B.

Certainly. The screw, with the addition of the lever, forms a very powerful machine, employed either for compression or to raise heavy weights. It is used by book-binders, to press the leaves of books together; it is used also in cyder and wine presses, in coining, and for a variety of other purposes.

All machines are composed of one or more of these six mechanical powers we have examined: I have but one more remark to make to you, relative to them, which is, that friction in a considerable degree diminishes their force, allowance must therefore always be made for it, in the construction of machinery.

CAROLINE.

By friction, do you mean one part of the machine rubbing against another part contiguous to it?

MRS. B.

Yes; friction is the resistance which bodies meet with in rubbing against each other; there is no such thing as perfect smoothness or evenness in nature: polished metals, though they wear that appearance, more than any other bodies, are far from really possessing it; and their inequalities may frequently be perceived through a good magnifying glass. When, therefore, the surfaces of the two bodies, come into contact, the prominent parts of the one will often

G 4

fall into the hollow parts of the other, and occasion more or less resistance to motion.

CAROLINE.

But if a machine is made of polished metal, as a watch for instance, the friction, must be very trifling?

MRS. B.

In proportion as the surfaces of bodies are well polished, the friction is doubtless diminished; but it is always considerable, and it is usually computed to destroy one-third of the power of a machine. Oil or grease is used to lessen friction: it acts as a polish by filling up the cavities of the rubbing surfaces, and thus making them slide more easily over each other.

CAROLINE.

Is it for this reason that wheels are greased, and the locks and hinges of doors oiled?

MRS. B.

Yes; in these instances the contact of the rubbing surfaces is so close, and the rubbing so continual, that notwithstanding their being polished and oiled, a considerable degree of friction is produced.

There are two kinds of friction; the one occasioned by the sliding of the flat surface of a body, the other by the rolling of a circular body: the friction resulting from the first is much the most considerable, for great force is required to enable the sliding body to overcome the resistance which the asperities of the surfaces in contact oppose to its motion, and it must be either lifted over, or break through them; whilst, in the other kind of friction, the rough parts roll over each other with comparative facility; hence it is, that wheels are often used for the sole purpose of diminishing the resistance of friction.

EMILY.

This is one of the advantages of carriage-wheels; is it not?

MRS. B.

Yes; and the larger the circumference of the wheel the more readily it can overcome any considerable obstacles, such as stones, or inequalities in the road. When, in descending a steep hill, we fasten one of the wheels, we decrease the velocity of the carriage, by increasing the friction.

CAROLINE.

That is to say, by converting the rolling friction into the dragging friction. And when you had casters put to the legs of the table, in order to move it more easily, you changed the dragging into the rolling friction.

G 5

#### MRS. B.

There is another circumstance which we have already noticed, as diminishing the motion of bodies, and which greatly affects the power of machines. This is the resistance of the medium, in which a machine is worked. All fluids, whether of the nature of air, or of water, are called mediums; and their resistance is proportioned to their density; for the more matter a body contains, the greater the resistance it will oppose to the motion of another body striking against it.

#### EMILY.

It would then be much more difficult to work a machine under water than in the air?

#### MRS. B.

Certainly, if a machine could be worked *in vacuo*, and without friction, it would be perfect; but this is unattainable; a considerable reduction of power must therefore be allowed for the resistance of the air.

We shall here conclude our observations on the mechanical powers. At our next meeting I shall endeavour to give you an explanation of the motion of the heavenly bodies.

# CONVERSATION VI.

## CAUSES OF THE EARTH'S ANNUAL MOTION.

OF THE PLANETS, AND THEIR MOTION. — OF THE
DIURNAL MOTION OF THE EARTH AND PLANETS.

CAROLINE.

I AM come to you to-day quite elated with the spirit of opposition, Mrs. B.; for I have discovered such a powerful objection to your theory of attraction, that I doubt whether even your conjuror Newton, with his magic wand of attraction, will be able to dispel it.

MRS. B,

Well my dear, pray what is this weighty objection?

CAROLINE.

You say that bodies attract in proportion to the

G 6

quantity of matter they contain, now we all know the sun to be much larger than the earth: why, therefore, does it not attract the earth; you will not, I suppose, pretend to say that we are falling towards the sun?

### EMILY.

However plausible your objection appears, Caroline, I think you place too much reliance upon it: when any one has given such convincing proofs of sagacity and wisdom as Sir Isaac Newton, when we find that his opinions are universally received and adopted, is it to be expected that any objection we can advance should overturn them?

### CAROLINE.

Yet I confess that I am not inclined to yield implicit faith even to opinions of the great Newton; for what purpose are we endowed with reason, if we are denied the privilege of making use of it, by judging for ourselves?

### MRS. B.

It is reason itself which teaches us, that when we, novices in science, start objections to theories established by men of acknowledged wisdom, we should be diffident rather of our own than of their opinion. I am far from wishing to lay the least restraint on your questions; you cannot be better convinced of

PLATE VI.

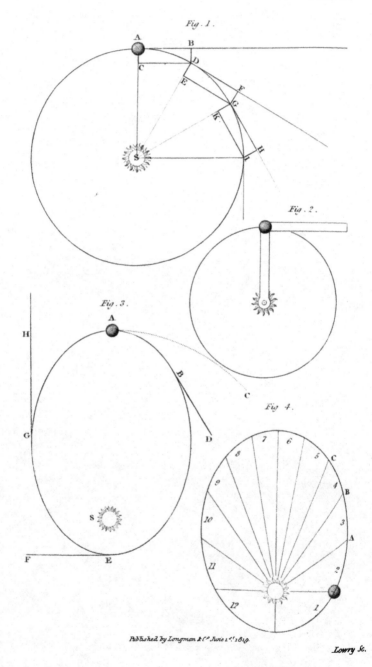

Fig. 1.

Fig. 2.

Fig. 3.

Fig. 4.

Published by Longman & Co. June 1.ˢᵗ 1819.

Lowry Sc.

the truth of a system, than by finding that it resists all your attacks, but I would advise you not to advance your objections with so much confidence, in order that the discovery of their fallacy may be attended with less mortification.   In answer to that you have just proposed, I can only say, that the earth really is attracted by the sun.

CAROLINE.

Take care at least that we are not consumed by him, Mrs. B.

MRS. B.

We are in no danger; but our magician Newton, as you are pleased to call him, cannot extricate himself from this difficulty without the aid of some cabalistical figures, which I must draw for him.

Let us suppose the earth, at its creation, to have been projected forwards into universal space; we know that if no obstacle impeded its course it would proceed in the same direction, and with a uniform velocity for ever.   In fig. 1. plate VI., A represents the earth, and S the sun.   We shall suppose the earth to be arrived at the point in which it is represented in the figure, having a velocity which would carry it on to B in the space of one month; whilst the sun's attraction would bring it to C in the same space of time.   Observe that the two forces of projection and attraction do not act in opposition, but perpendicularly, or at a right angle to each other. Can you tell me now, how the earth will move?

I recollect your teaching us that a body acted upon by two forces perpendicular to each other would move in the diagonal of a parallelogram; if, therefore, I complete the parallelogram by drawing the lines C D, B D, the earth will move in the diagonal A D.

A ball struck by two forces acting perpendicularly to each other, it is true, moves in the diagonal of a parallelogram; but you must observe that the force of attraction is continually acting upon our terrestrial ball, and producing an incessant deviation from its course in a right line, which converts it into that of a curve line; every point of which may be considered as constituting the diagonal of an infinitely small parallelogram.

Let us detain the earth a moment at the point D, and consider how it will be affected by the combined action of the two forces in its new situation. It still retains its tendency to fly off in a straight line; but a straight line would now carry it away to F, whilst the sun would attract it in the direction D S; how then will it proceed?

7

EMILY.

It will go on in a curve line in a direction between that of the two forces.

MRS. B.

In order to know exactly what course the earth will follow, draw another parallelogram similar to the first, in which the line D F describes the force of projection, and the line D S, that of attraction; and you will find that the earth will proceed in the curve line D G.

CAROLINE.

You must now allow me to draw a parallelogram, Mrs. B. Let me consider in what direction will the force of projection now impel the earth.

MRS. B.

First draw a line from the earth to the sun representing the force of attraction; then describe the force of projection at a right angle to it.

CAROLINE.

The earth will then move in the curve G I, of the parallelogram G H I K.

MRS. B.

You recollect that a body acted upon by two forces, moves through a diagonal in the same time that it would have moved through one of the sides of the parallelogram, were it acted upon by one force only. The earth has passed through the diagonals of these three parallelograms in the space of three months, and has performed one quarter of a circle; and on the same principle it will go on till it has completed the whole of the circle. It will then recommence a course, which it has pursued ever since it first issued from the hand of its Creator, and which there is every reason to suppose it will continue to follow, as long as it remains in existence.

### EMILY.

What a grand and beautiful effect, resulting from so simple a cause!

### CAROLINE.

It affords an example, on a magnificent scale, of the circular motion which you taught us in mechanics. The attraction of the sun is the centripetal force, which confines the earth to a centre; and the impulse of projection the centrifugal force, which impels the earth to quit the sun and fly off in a tangent.

### MRS. B.

Exactly so. A simple mode of illustrating the effect of these combined forces on the earth, is to cut a slip of card in the form of a right angle, (fig. 2. plate VI.) to describe a small circle at the angular point representing the earth, and to fasten the extremity of one of the legs of the angle to a fixed point, which we shall consider as the sun. Thus situated, the angle will represent both the centrifugal and centripetal forces; and if you draw it round the fixed point, you will see how the direction of the centrifugal force varies, constantly forming a tangent to the circle in which the earth moves, as it is constantly at a right angle with the centripetal force.

### EMILY.

The earth, then, gravitates towards the sun without the slightest danger either of approaching nearer or receding further from it. How admirably this is contrived! If the two forces which produce this circular motion had not been so accurately adjusted, one would ultimately have prevailed over the other, and we should either have approached so near the sun as to have been burnt, or have receded so far from it as to have been frozen.

### MRS. B.

What will you say, my dear, when I tell you, that these two forces are not, in fact, so proportioned as to produce circular motion in the earth?

You must explain to us, at least, in what manner we avoid the threatened destruction.

Let us suppose that when the earth is at A,
(fig. 3.) its projectile force should not have given it
a velocity sufficient to counterbalance that of gravity, so as to enable these powers conjointly to carry
it round the sun in a circle; the earth, instead of
describing the line A C, as in the former figure,
will approach nearer the sun in the line A B.

Under these circumstances, I see not what is to
prevent our approaching nearer and nearer the sun
till we fall into it; for its attraction increases as we
advance towards it, and produces an accelerated
velocity in the earth which increases the danger.

And there is yet another danger, of which you
are not aware. Observe, that as the earth approaches the sun, the direction of its projectile
force is no longer perpendicular to that of attraction, but inclines more nearly to it. When the
earth reaches that part of its orbit at B, the force of
projection would carry it to D, which brings it
nearer the sun instead of bearing it away from it.

EMILY.

If, then, we are driven by one power and drawn by the other to this centre of destruction, how is it possible for us to escape?

MRS. B.

A little patience, and you will find that we are not without resource. The earth continues approaching the sun with a uniformly increasing accelerated motion, till it reaches the point E: in what direction will the projectile force now impel it?

EMILY.

In the direction E F. Here then the two forces act perpendicularly to each other, and the earth is situated just as it was in the preceding figure; therefore, from this point, it should revolve round the sun in a circle.

MRS. B.

No, all the circumstances do not agree. In motion round a centre, you recollect that the centrifugal force increases with the velocity of the body, or in other words, the quicker it moves the stronger is its tendency to fly off in a right line. When the earth, therefore, arrives at E, its accelerated motion will have so far increased its velocity, and consequently its centrifugal force, that the latter will

prevail over the force of attraction, and drag the
earth away from the sun till it reaches G.

CAROLINE.

It is thus, then, that we escape from the dangerous vicinity of the sun; and in proportion as we
recede from it, the force of its attraction, and, consequently, the velocity of the earth's motion, are diminished.

MRS. B.

Yes. From G the direction of projection is towards H, that of attraction towards S, and the
earth proceeds between them with a uniformly retarded motion, till it has completed its revolution.
Thus you see, that the earth travels round the sun,
not in a circle, but an ellipsis, of which the sun
occupies one of the *foci*; and that in its course the
earth alternately approaches and recedes from it,
without any danger of being either swallowed up,
or of being entirely carried away from it.

CAROLINE.

And I observe, that what I apprehended to be a
dangerous irregularity, is the means by which the
most perfect order and harmony are produced!

EMILY.

The earth travels, then, at a very unequal rate,

its velocity being accelerated as it approaches the
sun, and retarded as it recedes from it.

MRS. B.

It is mathematically demonstrable, that, in moving round a point towards which it is attracted, a body
passes over equal areas in equal times.  The whole
of the space contained within the earth's orbit, is, in
fig. 4., divided into a number of areas, or spaces, 1,
2, 3, 4, &c. all of which are of equal dimensions,
though of very different forms; some of them, you see,
are long and narrow, others broad and short; but
they each of them contain an equal quantity of space.
An imaginary line drawn from the centre of the
earth to that of the sun, and keeping pace with the
earth in its revolution, passes over equal areas in
equal times; that is to say, if it is a month going
from A to B, it will be a month going from B to
C, and another from C to E, and so on.

CAROLINE.

What long journeys the earth has to perform in
the course of a month, in one part of her orbit, and
how short they are in the other part!

MRS. B.

The inequality is not so considerable as appears
in this figure; for the earth's orbit is not so eccen-
tric as it is there described; and, in reality, differs

but little from a circle : that part of the earth's orbit nearest the sun is called its *perihelion,* that part
most distant from the sun its *aphelion ;* and the
earth is above three millions of miles nearer the sun
at its perihelion than at its aphelion.

### EMILY.

I think I can trace a consequence from these different situations of the earth; is it not the cause
of summer and winter?

### MRS. B.

On the contrary; during the height of summer,
the earth is in that part of its orbit which is most
distant from the sun, and it is during the severity
of winter, that it approaches nearest to it.

### EMILY.

That is very extraordinary; and how then do
you account for the heat being greatest, when we
are most distant from the sun?

### MRS. B.

The difference of the earth's distance from the
sun in summer and winter, when compared with its
total distance from the sun, is but inconsiderable.
The earth, it is true, is above three millions of miles
nearer the sun in winter than in summer; but that
distance, however great it at first appears, sinks into

insignificance in comparison of 95 millions of miles, which is our mean distance from the sun. The change of temperature, arising from this difference, would scarcely be sensible; were it not completely overpowered by other causes which produce the variations of the seasons; but these I shall defer explaining, till we have made some further observations on the heavenly bodies.

CAROLINE.

And should not the sun appear smaller in summer, when it is so much further from us?

MRS. B.

It actually does, when accurately measured; but the apparent difference in size, is, I believe, not perceptible to the naked eye.

EMILY.

Then, since the earth moves with greatest velocity in that part of its orbit nearest the sun, it must have completed its journey through one half of its orbit in a shorter time than the other half?

MRS. B.

Yes, it is about seven days longer performing the summer-half of its orbit, than the winter-half.

The revolution of all the planets round the sun

is the result of the same causes, and is performed in the same manner as that of the earth.

CAROLINE.

Pray what are the planets?

MRS. B.

They are those celestial bodies, which revolve like our earth about the sun; they are supposed to resemble the earth also in many other respects; and we are led by analogy to suppose them to be inhabited worlds.

CAROLINE.

I have heard so; but do you not think such an opinion too great a stretch of the imagination?

MRS. B.

Some of the planets are proved to be larger than the earth; it is only their immense distance from us, which renders their apparent dimensions so small. Now if we consider them as enormous globes, instead of small twinkling spots, we shall be led to suppose, that the Almighty would not have created them merely for the purpose of giving us a little light in the night, as it was formerly imagined, and we should find it more consistent with our ideas of the Divine wisdom and beneficence,

to suppose that these celestial bodies, should be
created for the habitation of beings, who are, like
us, blessed by His providence. Both in a moral as
well as a physical point of view, it appears to me
more rational to consider the planets as worlds revolving round the sun; and the fixed stars as other
suns, each of them attended by their respective system of planets, to which they impart their influence.
We have brought our telescopes to such a degree
of perfection, that from the appearances which the
moon exhibits when seen through them, we have very
good reason to conclude, that it is a habitable
globe, for though it is true, that we cannot discern
its towns and people, we can plainly perceive its
mountains and valleys; and some astronomers have
gone so far as to imagine they discovered volcanos.

EMILY.

If the fixed stars are suns, with planets revolving round them, why should we not see those planets as well as their suns?

MRS. B.

In the first place, we conclude that the planets
of other systems, (like those of our own.) are much
smaller than the suns which give them light; therefore at so great a distance as to make the suns appear
like fixed stars, the planets would be quite invisible.
Secondly, the light of the planets being only re-

H

flected light, is much more feeble than that of the fixed stars. There is exactly the same difference as between the light of the sun and that of the moon; the first being a fixed star, the second a planet.

### EMILY.

But if the planets are worlds like our earth, they are dark bodies; and instead of shining by night, we should see them only by day-light. And why do we not see the fixed stars also by day-light?

### MRS. B.

Both for the same reason;—their light is so faint, compared to that of our sun reflected by the atmosphere, that it is entirely effaced by it: the light emitted by the fixed stars may probably be as strong as that of our sun, at an equal distance; but being so much more remote, it is diffused over a greater space, and is consequently proportionally weakened.

### CAROLINE.

True; I can see much better by the light of a candle that is near me, than by that of one at a great distance. But I do not understand what makes the planets shine?

### MRS. B.

What is it that makes the steel buttons on your brother's coat shine?

16

CAROLINE.

The sun.   But if it was the sun which made the planets shine, we should see them in the day-time, when the sun shone upon them; or if the faintness of their light prevented our seeing them in the day, we should not see them at all, for the sun cannot shine upon them in the night.

MRS. B.

There you are in error.   But in order to explain this to you, I must first make you acquainted with the various motions of the planets.

You know, that according to the laws of attraction, the planets belonging to our system all gravitate towards the sun; and that this force, combined with that of projection, will occasion their revolution round the sun, in orbits more or less elliptical, according to the proportion which these two forces bear to each other.

But the planets have also another motion: they revolve upon their axes.   The axis of a planet is an imaginary line which passes through its centre, and on which it turns; and it is this motion which produces day and night.   With that side of the planet facing the sun, it is day; and with the opposite side, which remains in darkness, it is night. Our earth, which we consider as a planet, is 24 hours in performing one revolution on its axis: in that period of time, therefore, we have a day and a night; hence this revolution is called the earth's diurnal or daily motion; and it is this revolution of the earth from west to east which produces an apparent motion of the sun, moon, and stars in a contrary direction.

Let us now suppose ourselves to be beings, independent of any planet, travelling in the skies, and looking upon the earth, in the same point of view as upon the other planets.

### CAROLINE.

It is not flattering to us, its inhabitants, to see it make so insignificant an appearance.

### MRS. B.

To those who are accustomed to contemplate it in this light, it never appears more glorious. We are taught by science to distrust appearances; and instead of considering the planets as little stars, we look upon them either as brilliant suns or habitable worlds, and we consider the whole together as forming one vast and magnificent system, worthy of the Divine hand by which it was created.

### EMILY.

I can scarcely conceive the idea of this immensity of creation; it seems too sublime for our imagination:— and to think that the goodness of Providence extends over millions of worlds throughout a boundless universe — Ah! Mrs. B., it is we only who become trifling and insignificant beings in so magnificent a creation !

This idea should teach us humility, but without producing despondency.  The same Almighty hand which guides these countless worlds in their undeviating course, conducts with equal perfection the blood as it circulates through the veins of a fly, and opens the eye of the insect to behold His wonders. Notwithstanding this immense scale of creation, therefore, we need not fear to be disregarded or forgotten.

But to return to our station in the skies.  We were, if you recollect, viewing the earth at a great distance, in appearance a little star, one side illumined by the sun, the other in obscurity.   But would you believe it, Caroline, many of the inhabitants of this little star imagine that when that part which they inhabit is turned from the sun, darkness prevails throughout the universe, merely because it is night with them; whilst, in reality, the sun never ceases to shine upon every planet.   When, therefore, these little ignorant beings look around them during their night, and behold all the stars shining, they cannot imagine why the planets, which are dark bodies, should shine, concluding, that since the sun does not illumine themselves, the whole universe must be in darkness.

CAROLINE.

I confess that I was one of these ignorant people; but I am now very sensible of the absurdity of such an idea.    To the inhabitants of the other planets, then, we must appear as a little star?

MRS. B.

Yes, to those which revolve round our sun; for since those which may belong to other systems (and whose existence is only hypothetical,) are invisible to us, it is probable, that we also are invisible to them.

EMILY.

But they may see our sun as we do theirs, in appearance a fixed star?

MRS. B.

No doubt; if the beings who inhabit those planets are endowed with senses similar to ours.    By the same rule, we must appear as a moon, to the inhabitants of our moon; but on a larger scale, as the surface of the earth is about thirteen times as large as that of the moon.

EMILY.

The moon, Mrs. B., appears to move in a different direction, and in a different manner from the stars?

MRS. B.

I shall defer the explanation of the motion of the moon, till our next interview, as it would prolong our present lesson too much.

# CONVERSATION VII.

## ON THE PLANETS.

OF THE SATELLITES OR MOONS. — GRAVITY DIMI-
NISHES AS THE SQUARE OF THE DISTANCE. —
OF THE SOLAR SYSTEM. — OF COMETS. — CON-
STELLATIONS, SIGNS OF THE ZODIAC. — OF
COPERNICUS, NEWTON, &c.

MRS. B.

THE planets are distinguished into primary and
secondary. Those which revolve immediately
about the sun are called primary. Many of these are
attended in their course by smaller planets, which
revolve round them: these are called secondary
planets, satellites, or moons. Such is our moon,
which accompanies the earth, and is carried with
it round the sun.

### EMILY.

How then can you reconcile the motion of the secondary planets to the laws of gravitation; for the sun is much larger than any of the primary planets; and is not the power of gravity proportional to the quantity of matter?

### CAROLINE.

Perhaps the sun, though much larger, may be less dense than the planets. Fire you know is very light, and it may contain but little matter, though of great magnitude.

### MRS. B.

We do not know of what kind of matter the sun is made; but we may be certain, that since it is the general centre of attraction of our system of planets, it must be the body which contains the greatest quantity of matter in that system.

You must recollect, that the force of attraction is not only proportional to the quantity of matter, but to the degree of proximity of the attractive body: this power is weakened by being diffused, and diminishes as the squares of the distances increase. The square is the product of a number multiplied by itself; so that a planet situated at twice the distance at which we are from the sun would gravitate four times less than we do; for the product of two multiplied by itself is four.

H 5

**CAROLINE.**

Then the more distant planets move slower in
their orbits; for their projectile force must be proportioned to that of attraction? But I do not see
how this accounts for the motion of the secondary
round the primary planets, in preference to the
sun?

**EMILY.**

Is it not because the vicinity of the primary
planets renders their attraction stronger than that
of the sun?

**MRS. B.**

Exactly so. But since the attraction between
bodies is mutual, the primary planets are also
attracted by the satellites, which revolve round
them. The moon attracts the earth, as well as the
earth the moon; but as the latter is the smaller
body, her attraction is proportionally less; therefore neither the earth revolves round the moon, nor
the moon round the earth; but they both revolve
round a point, which is their common centre of
gravity, and which is as much nearer the earth
than the moon, as the gravity of the former exceeds
that of the latter.

**EMILY.**

Yes, I recollect your saying, that if two bodies

were fastened together by a wire or bar, their common centre of gravity would be in the middle of the bar, provided the bodies were of equal weight; and if they differed in weight, it would be nearer the larger body. If then the earth and moon had no projectile force which prevented their mutual attraction from bringing them together, they would meet at their common centre of gravity.

CAROLINE.

The earth then has a great variety of motions. it revolves round the sun, upon its axis, and round the point towards which the moon attracts it.

MRS. B.

Just so; and this is the case with every planet which is attended by satellites. The complicated effect of this variety of motions, produces certain irregularities, which, however, it is not necessary to notice at present.

The planets act on the sun in the same manner as they are themselves acted on by their satellites; for attraction, you must remember, is always mutual; but the gravity of the planets (even when taken collectively) is so trifling compared with that of the sun, that they do not cause the latter to move so much as one half of his diameter. The planets do not, therefore, revolve round the centre of the

H 6

sun, but round a point at a small distance from its centre, about which the sun also revolves.

I thought the sun had no motion?

You were mistaken; for, besides that which I have just mentioned, which is indeed very inconsiderable, he revolves on his axis; this motion is ascertained by observing certain spots which disappear, and reappear regularly at stated times.

A planet has frequently been pointed out to me in the heavens; but I could not perceive that its motion differed from that of the fixed stars, which only appear to move.

The great distance of the planets renders their motion apparently so slow, that the eye is not sensible of their progress in their orbit, unless we watch them for some considerable length of time: in different seasons they appear in different parts of the heavens. The most accurate idea I can give you of the situation and motion of the planets, will be by the examination of this diagram, (Plate VII. fig. 1.) repre-

PLATE VII

Fig. 1

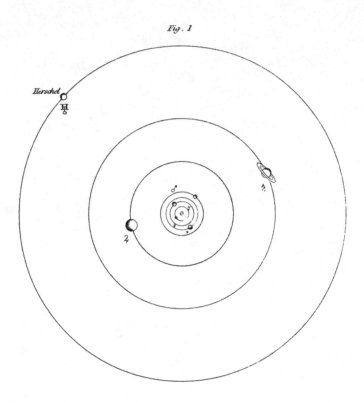

Herschel

Fig. 2.

Jupiter

Saturn

Mercury  Mars  Venus  Earth  Herschel

Moon

Published by Longman & Co. June 1.ˢᵗ 1849.

Lowry Sc.

senting the solar system, in which you will find every planet with its orbit delineated.

EMILY.

But the orbits here are all circular, and you said that they were elliptical. The planets appear too, to be moving round the centre of the sun; whilst you told us, that they moved round a point at a little distance from thence.

MRS. B.

The orbits of the planets are so nearly circular, and the common centre of gravity of the solar system so near the centre of the sun, that these deviations are scarcely worth observing. The dimensions of the planets, in their true proportions, you will find delineated in fig. 2.

Mercury is the planet nearest the sun; his orbit is consequently contained within ours; but his vicinity to the sun, occasions his being nearly lost in the brilliancy of his rays; and when we see the sun, he is so dazzling, that very accurate observations cannot be made upon Mercury. He performs his revolution round the sun in about 87 days, which is consequently the length of his year. The time of his rotation on his axis is not known; his distance from the sun is computed to be 37 millions of miles, and his diameter 3180 miles. The heat of this planet is so great, that water cannot exist there, but in a state of vapour, and metals would be liquified.

CAROLINE.

Oh, what a dreadful climate!

MRS. B.

Though we could not live there, it may be perfectly adapted to other beings destined to inhabit it.

Venus, the next in the order of planets, is 68 millions of miles from the sun: she revolves about her axis in 23 hours and 21 minutes, and goes round the sun in 244 days 17 hours. The orbit of Venus is also within ours; during one half of her course in it, we see her before sun-rise, and she is called the morning star; in the other part of her orbit, she rises later than the sun.

CAROLINE.

In that case, we cannot see her, for she must rise in the day time?

MRS. B.

True; but when she rises later than the sun, she also sets later; so that we perceive her approaching the horizon after sun-set: she is then called Hesperus, or the evening star. Do you recollect those beautiful lines of Milton:

Now came still evening on, and twilight gray
Had in her sober livery all things clad;

Silence accompanied; for beast and bird,
They to their grassy couch, these to their nests
Were slunk, all but the wakeful nightingale;
She all night long her amorous descant sung;
Silence was pleas'd: now glow'd the firmament
With living saphirs: Hesperus, that led
The starry host, rode brightest, till the moon
Rising in clouded majesty, at length
Apparent queen unveil'd her peerless light,
And o'er the dark her silver mantle threw.

The planet next to Venus is the Earth, of which we shall soon speak at full length. At present I shall only observe, that we are 95 millions of miles distant from the sun, that we perform our annual revolution in 365 days 5 hours and 49 minutes; and are attended in our course by a single moon.

Next follows Mars. He can never come between us and the sun, like Mercury and Venus; his motion is, however, very perceptible, as he may be traced to different situations in the heavens; his distance from the sun is 144 millions of miles; he turns round his axis in 24 hours and 39 minutes; and he performs his annual revolution, in about 687 of our days: his diameter is 4120 miles. Then follow four very small planets, Juno, Ceres, Pallas, and Vesta, which have been recently discovered, but whose dimensions and distances from the sun have not been very accurately ascertained.

Jupiter is next in order: this is the largest of all the planets. He is about 490 millions of miles from the sun, and completes his annual period in nearly twelve of our years. He turns round his axis in about ten hours. He is above 1200 times as big as our earth; his diameter being 86,000 miles. The respective proportions of the planets cannot, therefore, you see, be conveniently delineated in a diagram. He is attended by four moons.

The next planet is Saturn, whose distance from the sun is about 900 millions of miles; his diurnal rotation is performed in 10 hours and a quarter: — his annual revolution in nearly 30 of our years. His diameter is 79,000 miles. This planet is surrounded by a luminous ring, the nature of which, astronomers are much at a loss to conjecture; he has seven moons. Lastly, we observe the Georgium Sidus, discovered by Dr. Herschel, and which is attended by six moons.

CAROLINE.

How charming it must be in the distant planets, to see several moons shining at the same time; I think I should like to be an inhabitant of Jupiter or Saturn.

MRS. B.

Not long, I believe. Consider what extreme cold must prevail in a planet, situated as Saturn is, at nearly ten times the distance at which we are from the sun. Then his numerous moons are far from making so splendid an appearance as ours; for they can reflect only the light which they receive from the sun; and both light and heat decrease in the same ratio or proporton to the distances as gravity. Can you tell me now how much more light we enjoy than Saturn.

CAROLINE.

The square of ten, is a hundred; therefore, Saturn has a hundred times less — or to answer your question exactly, we have a hundred times more light and heat than Saturn — this certainly does not increase my wish to become one of the poor wretches who inhabit that planet.

MRS. B.

May not the inhabitants of Mercury, with equal plausibility, pity us, for the insupportable coldness of our situation; and those of Jupiter and Saturn for our intolerable heat? The Almighty Power which created these planets, and placed them in their several orbits, has no doubt peopled them with beings whose bodies are adapted to the various temperatures and elements in which they are situated. If we judge from the analogy of our own earth, or from that of the great and universal beneficence of Providence, we must conclude this to be the case.

CAROLINE.

Are not comets also supposed to be planets?

Yes, they are; for by the re-appearance of some
of them, at stated times, they are known to revolve
round the sun, but in orbits so extremely excentric,
that they disappear for a great number of years. If
they are inhabited, it must be by a species of beings
very different, not only from the inhabitants of this,
but from those of any of the other planets, as they
must experience the greatest vicissitudes of heat and
cold; one part of their orbit being so near the sun,
that their heat, when there, is computed to be greater
than that of red-hot iron; in this part of its orbit,
the comet emits a luminous vapour, called the tail,
which it gradually loses as it recedes from the sun;
and the comet itself totally disappears from our
sight, in the more distant parts of its orbit, which
extends considerably beyond that of the furthest
planet.

The number of comets belonging to our system,
cannot be ascertained, as some of them are whole
centuries before they make their re-appearance.
The number that are known by their regular re-
appearance is only three.

Pray, Mrs. B., what are the constellations?

They are the fixed stars, which the ancients, in

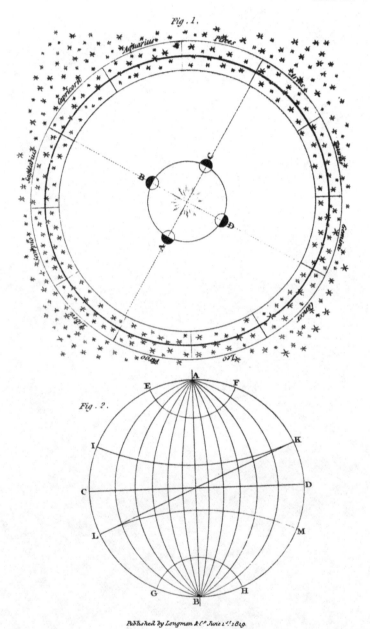

PLATE VIII.

Fig. 1.

Fig. 2.

Published by Longman & C.º June 1.ˢᵗ 1829.

Lowry Sc.

order to recognise them, formed into groups, and
gave the names of the figures, which you find delineated on the celestial globe. In order to show
their proper situations in the heavens, they should
be painted on the internal surface of a hollow
sphere, from the centre of which you should view
them; you would then behold them, as they appear
to be situated in the heavens. The twelve constellations, called the signs of the zodiac, are those
which are so situated, that the earth in its annual
revolution passes directly between them and the
sun. Their names are Aries, Taurus, Gemini,
Cancer, Leo, Virgo, Libra, Scorpio, Sagittarius,
Capricornus, Aquarius, Pisces; the whole occupying a complete circle, or broad belt, in the
heavens, called the zodiac. (Plate VIII. fig. 1.)
Hence, a right line drawn from the earth, and
passing through the sun, would reach one of these
constellations, and the sun is said to be in that constellation at which the line terminates : thus, when
the earth is at A, the sun would appear to be in
the constellation or sign Aries ; when the earth is at
B, the sun would appear in Cancer ; when the earth
was at C, the sun would be in Libra ; and when
the earth was at D, the sun would be in Capricorn.
This circle, in which the sun thus appears to move,
and which passes through the middle of the zodiac,
is called the ecliptic,

But many of the stars in these constellations appear beyond the zodiac.

MRS. B.

We have no means of ascertaining the distance of the fixed stars. When, therefore, they are said to be in the zodiac, it is merely implied, that they are situated in that direction, and that they shine upon us through that portion of the heavens, which we call the zodiac.

EMILY.

But are not those large bright stars, which are called stars of the first magnitude, nearer to us, than those small ones which we can scarcely discern?

MRS. B.

It may be so; or the difference of size and brilliancy of the stars may proceed from their difference of dimensions; this is a point which astronomers are not enabled to determine. Considering them as suns, I see no reason why different suns should not vary in dimensions, as well as the planets belonging to them.

EMILY.

What a wonderful and beautiful system this is,

and how astonishing to think that every fixed star may probably be attended by a similar train of planets !

CAROLINE.

You will accuse me of being very incredulous, but I cannot help still entertaining some doubts, and fearing that there is more beauty than truth in this system. It certainly may be so; but there does not appear to me to be sufficient evidence to prove it. It seems so plain and obvious that the earth is motionless, and that the sun and stars revolve round it; — your solar system, you must allow, is directly in opposition to the evidence of our senses.

MRS. B.

Our senses so often mislead us, that we should not place implicit reliance upon them.

CAROLINE.

On what then can we rely, for do we not receive all our ideas through the medium of our senses?

MRS. B.

It is true, that they are our primary source of knowledge; but the mind has the power of reflecting, judging, and deciding upon the ideas received by the organs of sense. This faculty, which we call reason, has frequently proved to us, that our senses

are liable to err.   If you  have  ever  sailed  on  the
water,  with  a  very  steady  breeze,  you  must have
seen the houses, trees, and every object move, while
you were sailing.

I remember thinking so, when I was very young;
but I now know that their motion is only apparent.
It is true that my reason, in this case, corrects the
error of my sight.

It teaches you,  that the apparent motion of the
objects on shore, proceeds from your being yourself
moving,  and that you are not sensible of your own
motion, because you meet with no resistance.   It is
only when some obstacle impedes our motion,  that
we are conscious of moving; and if you were to
close your eyes when you were sailing on calm
water,  with a steady wind, you would not perceive
that you moved, for you could not feel it, and you
could see it only by observing the change of place
of the objects on shore.   So it is with the motion
of the earth ; every thing on its surface, and the air
that surrounds it, accompanies it in its revolution ;
it meets with no resistance: therefore, like the crew
of a vessel sailing with a fair wind,  in a calm sea,
we are insensible of our motion.

CAROLINE.

But the principal reason why the crew of a vessel in a calm sea do not perceive their motion, is, because they move exceedingly slowly; while the earth, you say, revolves with great velocity.

MRS. B.

It is not because they move slowly, but because they move steadily, and meet with no irregular resistances, that the crew of a vessel do not perceive their motion; for they would be equally insensible to it, with the strongest wind, provided it were steady, that they sailed with it, and that it did not agitate the water; but this last condition, you know, is not possible, for the wind will always produce waves which offer more or less resistance to the vessel, and then the motion becomes sensible, because it is unequal.

CAROLINE.

But, granting this, the crew of a vessel have a proof of their motion, though insensible, which the inhabitants of the earth cannot have, — the apparent motion of the objects on shore.

MRS. B.

Have we not a similar proof of the earth's motion, in the apparent motion of the sun and stars. Imagine the earth to be sailing round its axis, and successively passing by every star, which, like the objects on land, we suppose to be moving instead of ourselves.  I have heard it observed by an aerial traveller in a balloon, that the earth appears to sink beneath the balloon, instead of the balloon rising above the earth.

It is a law which we discover throughout nature and worthy of its great Author, that all its purposes are accomplished by the most simple means; and what reason have we to suppose this law infringed, in order that we may remain at rest, while the sun and stars move round us; their regular motions, which are explained by the laws of attraction on the first supposition, would be unintelligible on the last, and the order and harmony of the universe be destroyed.  Think what an immense circuit the sun and stars would make daily, were their apparent motions real.  We know many of them to be bodies more considerable than our earth; for our eyes vainly endeavour to persuade us, that they are little brilliants sparkling in the heavens, while science teaches us that they are immense spheres, whose apparent dimensions are diminished by distance. Why then should these enormous globes daily traverse such a prodigious space, merely to prevent the necessity of our earth's revolving on its axis?

CAROLINE.

I think I must now be convinced.  But you will, I hope, allow me a little time to familiarise myself to an idea so different from that which I have been accustomed to entertain. And pray, at what rate do we move?

### MRS. B.

The motion produced by the revolution of the earth on its axis, is about eleven miles a minute, to an inhabitant of London.

### EMILY.

But does not every part of the earth move with the same velocity?

### MRS. B.

A moment's reflection would convince you of the contrary: a person at the equator must move quicker than one situated near the poles, since they both perform a revolution in 24 hours.

### EMILY.

True, the equator is farthest from the axis of motion. But in the earth's revolution round the sun, every part must move with equal velocity?

### MRS. B.

Yes, about a thousand miles a minute.

### CAROLINE.

How astonishing! — and that it should be pos-

I

sible for us to be insensible of such a rapid motion.
You would not tell me this sooner, Mrs. B., for fear
of increasing my incredulity.

Before the time of Newton, was not the earth
supposed to be in the centre of the system, and the
sun, moon, and stars to revolve round it?

### MRS. B.

This was the system of Ptolemy in ancient
times; but as long ago as the beginning of the six-
teenth century it was discarded, and the solar system,
such as I have shown you, was established by the
celebrated astronomer Copernicus, and is hence
called the Copernican system.   But the theory of
gravitation, the source from which this beautiful and
harmonious arrangement flows, we owe to the
powerful genius of Newton, who lived at a much
later period.

### EMILY.

It appears, indeed, far less difficult to trace by
observation the motion of the planets, than to
divine by what power they are impelled and guided.
I wonder how the idea of gravitation could first
have occurred to Sir Isaac Newton?

### MRS. B.

It is said to have been occasioned by a circumstance from which one should little have expected so grand a theory to have arisen. During the prevalence of the plague in the year 1665, Newton retired into the country to avoid the contagion: when sitting one day in his orchard, he observed an apple fall from a tree, and was led to consider what could be the cause which brought it to the ground.

CAROLINE.

If I dared to confess it, Mrs. B., I should say that such an enquiry indicated rather a deficiency than a superiority of intellect. I do not understand how any one can wonder at what is so natural and so common.

MRS. B.

It is the mark of superior genius to find matter for wonder, observation, and research, in circumstances which, to the ordinary mind, appear trivial, because they are common, and with which they are satisfied, because they are natural, without reflecting that nature is our grand field of observation, that within it is contained our whole store of knowledge; in a word, that to study the works of nature, is to learn to appreciate and admire the wisdom of God. Thus, it was the simple circumstance of the fall of an apple, which led to the discovery of the laws upon which the Copernican system is founded; and

I. 2

whatever credit this system had obtained before, it now rests upon a basis from which it cannot be shaken.

EMILY.

This was a most fortunate apple, and more worthy to be commemorated than all those that have been sung by the poets. The apple of discord for which the goddesses contended; the golden apples by which Atalanta won the race; nay, even the apple which William Tell shot from the head of his son, cannot be compared to this!

# CONVERSATION VIII.

## ON THE EARTH.

OF THE TERRESTRIAL GLOBE. — OF THE FIGURE
OF THE EARTH. — OF THE PENDULUM. — OF
THE VARIATION OF THE SEASONS, AND OF THE
LENGTH OF DAYS AND NIGHTS. — OF THE CAUSES
OF THE HEAT OF SUMMER. — OF SOLAR, SIDERIAL,
AND EQUAL OR MEAN TIME.

MRS. B.

As the earth is the planet in which we are the
most particularly interested, it is my intention,
this morning, to explain to you the effects resulting from its annual and diurnal motions; but
for this purpose it will be necessary to make you
acquainted with the terrestrial globe: you have not
either of you, I conclude, learnt the use of the
globes?

CAROLINE.

No; I once indeed learnt by heart the names of
the lines marked on the globe, but as I was informed

I 3

they were only imaginary divisions, they did not appear to me worthy of much attention, and were soon forgotten.

### MRS. B.

You supposed, then, that astronomers had been at the trouble of inventing a number of lines to little purpose. It will be impossible for me to explain to you the particular effects of the earth's motion, without your having acquired a knowledge of these lines : in Plate VIII. fig. 2. you will find them all delineated; and you must learn them perfectly if you wish to make any proficiency in astronomy.

### CAROLINE.

I was taught them at so early an age that I could not understand their meaning; and I have often heard you say that the only use of words was to convey ideas.

### MRS. B.

The names of these lines would have conveyed ideas of the figures they were designed to express, though the use of these figures might at that time have been too difficult for you to understand. Childhood is the season when impressions on the memory are most strongly and most easily made: it is the period at which a large stock of ideas should be treasured up, the application of which we may learn when the understanding is more developed. It is, I think, a very mistaken notion that children should be taught such things only, as they can perfectly understand. Had you been early made acquainted with the terms which relate to figure and motion, how much it would have facilitated your progress in natural philosophy. I have been obliged to confine myself to the most common and familiar expressions, in explaining the laws of nature, though I am convinced that appropriate and scientific terms would have conveyed more precise and accurate ideas; but I was afraid of not being understood.

EMILY.

You may depend upon our learning the names of these lines thoroughly, Mrs. B.; but before we commit them to memory, will you have the goodness to explain them to us?

MRS. B.

Most willingly. This globe, or sphere, represents the earth; the line which passes through its centre, and on which it turns, is called its axis; and the two extremities of the axis, A and B, are the poles, distinguished by the names of the north and the south pole. The circle CD, which divides the globe into two equal parts between the poles, is called the equator, or equinoctial line; that part of the globe to the north of the equator is the northern hemisphere; that part to the south of the equator, the southern hemisphere. The small circle E F, which surrounds the north pole, is called the arctic circle; that G H, which surrounds the south pole, the antarctic circle. There are two intermediate circles between, the polar circles and the equator; that to the north, I K, called the tropic of Cancer; that to the south, L M, called the tropic of Capricorn. Lastly, this circle, L K, which divides the globe into two equal parts, crossing the equator and extending northward as far as the tropic of Cancer, and southward as far as the tropic of Capricorn, is called the ecliptic. The delineation of the ecliptic on the terrestrial globe is not without danger of conveying false ideas; for the ecliptic (as I have before said) is an imaginary circle in the heavens passing through the middle of the zodiac, and situated in the plane of the earth's orbit.

CAROLINE.

I do not understand the meaning of the plane of the earth's orbit.

MRS. B.

A plane, or plain, is an even level surface. Let us suppose a smooth thin solid plain cutting the sun through the centre, extending out as far as the fixed stars, and terminating in a circle which passes through the middle of the zodiac; in this plane the

PLATE IX

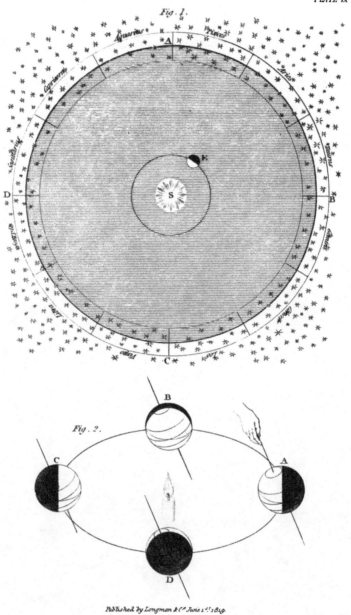

Fig. 1.

Fig. 2.

Published by Longman & Co. June 1.st 1819.

Lowry Sc.

earth would move in its revolution round the sun;
it is therefore called the plane of the earth's orbit,
and the circle in which this plane cuts the signs of
the zodiac is the ecliptic. Let the fig. 1. Plate IX.
represent such a plane, S the sun, E the earth with
its orbit, and A B C D the ecliptic passing through
the middle of the zodiac.

EMILY.

If the ecliptic relates only to the heavens, why is
it described upon the terrestrial globe?

MRS. B,

It is convenient for the demonstration of a variety
of problems in the use of the globes; and besides,
the obliquity of this circle to the equator is rendered more conspicuous by its being described on the
same globe; and the obliquity of the ecliptic shows
the inclination of the earth's axis to the plane of
its orbit. But to return to fig. 2, Plate VIII.

The spaces between the several parallel circles
on the terrestrial globe are called zones; that which
is comprehended between the tropics is distinguished by the name of the torrid zone; the spaces
which extend from the tropics to the polar circles,
the north and south temperate zones; and the spaces
contained within the polar circles, the frigid zones.

The several lines which, you observe, are drawn
from one pole to the other, cutting the equator at

I 5

right angles, are called meridians. When any one of these meridians is exactly opposite the sun it is mid-day, or twelve o'clock in the day, with all the places situated on that meridian; and, with the places situated on the opposite meridian, it is consequently midnight.

EMILY.

To places situated equally distant from these two meridians, it must then be six o'clock?

MRS. B.

Yes; if they are to the east of the sun's meridian it is six o'clock in the afternoon, because the sun will have previously passed over them; if to the west, it is six o'clock in the morning, and the sun will be proceeding towards that meridian.

Those circles which divide the globe into two equal parts, such as the equator and the ecliptic, are called greater circles; to distinguish them from those which divide it into two unequal parts, as the tropics and polar circles, which are called lesser circles. All circles are divided into 360 equal parts, called degrees, and degrees into 60 equal parts, called minutes. The diameter of a circle is a right line drawn across it, and passing through the centre; for instance, the boundary of this sphere is a circle, and its axis the diameter of that circle; the diameter is equal to a little less.

than one-third of the circumference. Can you tell me nearly how many degrees it contains?

CAROLINE.

It must be something less than one-third of 360 degrees, or nearly 120 degrees.

MRS. B.

Right; now Emily you may tell me exactly how many degrees are contained in a meridian?

EMILY.

A meridian, reaching from one pole to the other, is half a circle, and must therefore contain 180 degrees.

MRS. B.

Very well; and what number of degrees are there from the equator to the poles?

CAROLINE.

The equator being equally distant from either pole, that distance must be half of a meridian, or a quarter of the circumference of a circle, and contain 90 degrees.

MRS. B.

Besides the usual division of circles into degrees,

the ecliptic is divided into twelve equal parts, called
signs, which bear the name of the constellations
through which this circle passes in the heavens.
The degrees measured on the meridians from north
to south, or south to north, are called degrees of
latitude; those measured from east to west on the
equator, the ecliptic, or any of the lesser circles,
are called degrees of longitude; hence these circles
bear the name of longitudinal circles; they are also
called parallels of latitude.

EMILY.

The degrees of longitude must then vary in length
according to the dimensions of the circle on which
they are reckoned; those, for instance, at the polar
circles will be considerably smaller than those at
the equator?

MRS. B.

Certainly; since the degrees of circles of different
dimensions do not vary in number, they must necessarily vary in length.  The degrees of latitude,
you may observe, never vary in length; for the
meridians on which they are reckoned are all of the
same dimensions.

EMILY.

And of what length is a degree of latitude?

MRS. B.

Sixty geographical miles, which is equal to 69½ English statute miles.

EMILY.

The degrees of longitude at the equator must then be of the same dimensions?

MRS. B,

They would, were the earth a perfect sphere; but its form is not exactly spherical, being somewhat protuberant about the equator, and flattened towards the poles. This form is supposed to proceed from the superior action of the centrifugal power at the equator.

CAROLINE.

I thought I had understood the centrifugal force perfectly, but I do not comprehend its effect in this instance.

MRS. B.

You know that the revolution of the earth on its axis must give every particle a tendency to fly off from the centre, that this tendency is stronger or weaker in proportion to the velocity with which the particle moves; now a particle situated near one of the polar circles makes one rotation in the same space of time as a particle at the equator; the latter, therefore, having a much larger circle to describe, travels proportionally faster, consequently the centrifugal force is much stronger at the equator than at the polar circles : it gradually decreases as you leave the equator and approach the poles, where, as there is no rotatory motion, it entirely ceases. Supposing, therefore, the earth to have been originally in a fluid state, the particles in the torrid zone would recede much farther from the centre than those in the frigid zones; thus the polar regions would become flattened, and those about the equator elevated.

### CAROLINE.

I did not consider that the particles in the neighbourhood of the equator move with greater velocity than those about the poles; this was the reason I could not understand you.

### MRS. B.

You must be careful to remember, that those parts of a body which are farthest from the centre of motion must move with the greatest velocity : the axis of the earth is the centre of its diurnal motion, and the equatorial regions the parts most distant from the axis.

### CAROLINE.

My head then moves faster than my feet; and

upon the summit of a mountain we are carried round quicker than in a valley?

### MRS. B.

Certainly, your head is more distant from the centre of motion, than your feet; the mountain-top than the valley: and the more distant any part of a body is from the centre of motion, the larger is the circle it will describe, and the greater therefore must be its velocity.

### EMILY.

I have been reflecting, that if the earth is not a perfect circle.....

### MRS. B.

A sphere you mean, my dear; a circle is a round line, every part of which is equally distant from the centre; a sphere or globe is a round body, the surface of which is every where equally distant from the centre.

### EMILY.

If, then, the earth is not a perfect sphere, but prominent at the equator, and depressed at the poles, would not a body weigh heavier at the equator than at the poles? For the earth being thicker at the equator, the attraction of gravity perpendicularly downwards must be stronger.

Your reasoning has some plausibility, but I am sorry to be obliged to add, that it is quite erroneous; for the nearer any part of the surface of a body is to the centre of attraction, the more strongly it is attracted; because the most considerable quantity of matter is about that centre. In regard to its effects, you might consider the power of gravity, as that of a magnet placed at the centre of attraction.

EMILY.

But were you to penetrate deep into the earth, would gravity increase as you approached the centre?

MRS. B.

Certainly not; I am referring only to any situation on the surface of the earth. Were you to penetrate into the interior, the attraction of the parts above you would counteract that of the parts beneath you, and consequently diminish the power of gravity in proportion as you approached the centre; and if you reached that point, being equally attracted by the parts all around you, gravity would cease, and you would be without weight.

EMILY.

Bodies then should weigh less at the equator

than at the poles, since they are more distant from the centre of gravity in the former than in the latter situation?

MRS. B.

And this is really the case; but the difference of weight would be scarcely sensible, were it not augmented by another circumstance.

CAROLINE.

And what is this singular circumstance, which seems to disturb the laws of nature?

MRS. B.

One that you are well acquainted with, as conducing more to the preservation than the destruction of order, — the centrifugal force. This we have just observed to be stronger at the equator; and as it tends to drive bodies from the centre, it is necessarily opposed to, and must lessen the power of gravity, which attracts them towards the centre. We accordingly find that bodies weigh lightest at the equator, where the centrifugal force is greatest; and heaviest at the poles, where this power is least.

CAROLINE.

Has the experiment been made in these different situations?

186    ON THE EARTH.

MRS. B.

Lewis XIV., of France, sent philosophers both
to the equator and to Lapland for this purpose:
the severity of the climate, and obstruction of
the ice, has hitherto rendered every attempt to
reach the pole abortive; but the difference of
gravity at the equator and in Lapland is very perceptible.

CAROLINE.

Yet I do not comprehend, how the difference of
weight could be ascertained; for if the body under
trial decreased in weight, the weight which was opposed to it in the opposite scale must have diminished in the same proportion.  For instance, if a
pound of sugar did not weigh so heavy at the equator as at the poles, the leaden pound which served
to weigh it, would not be so heavy either; therefore
they would still balance each other, and the different force of gravity could not be ascertained by
this means.

MRS. B.

Your observation is perfectly just: the difference
of gravity of bodies situated at the poles and at
the equator cannot be ascertained by weighing
them; a pendulum was therefore used for that
purpose.

CAROLINE.

What, the pendulum of a clock ? how could that answer the purpose ?

MRS. B.

A pendulum consists of a line, or rod, to one end of which a weight is attached, and it is suspended by the other to a fixed point, about which it is made to vibrate. Without being put in motion, a pendulum, like a plumb line, hangs perpendicular to the general surface of the earth, by which it is attracted ; but, if you raise a pendulum, gravity will bring it back to its perpendicular position. It will, however, not remain stationary there, for the velocity it has received during its descent will impel it onwards, and it will rise on the opposite side to an equal height; from thence it is brought back by gravity, and again driven by the impulse of its velocity.

CAROLINE.

If so, the motion of a pendulum would be perpetual, and I thought you said, that there was no perpetual motion on the earth.

MRS. B.

The motion of a pendulum is opposed by the resistance of the air in which it vibrates, and by the friction of the part by which it is suspended : were it possible to remove these obstacles, the motion of a pendulum would be perpetual, and its vibrations perfectly regular; being of equal distances, and performed in equal times.

### EMILY.

That is the natural result of the uniformity of the power which produces these vibrations, for the force of gravity being always the same, the velocity of the pendulum must consequently be uniform.

### CAROLINE.

No, Emily, you are mistaken; the cause is not always uniform, and therefore the effect will not be so either. I have discovered it, Mrs. B.; since the force of gravity is less at the equator than at the pole , the vibrations of the pendulum will be slower at the equator than at the poles.

### MRS. B.

You are perfectly right, Caroline; it was by this means that the difference of gravity was discovered, and the true figure of the earth ascertained.

### EMILY.

But how do they contrive to regulate their time in the equatorial and polar regions? for, since in this part of the earth the pendulum of a clock vibrates exactly once in a second, if it vibrates faster at the poles and slower at the equator, the inhabitants must regulate their clocks in a different manner from ours.

#### MRS. B.

The only alteration required is to lengthen the pendulum in one case, and to shorten it in the other; for the velocity of the vibrations of a pendulum depends on its length; and when it is said, that a pendulum vibrates quicker at the pole than at the equator, it is supposing it to be of the same length. A pendulum which vibrates a second in this latitude is $36\frac{1}{2}$ inches long. In order to vibrate at the equator in the same space of time, it must be lengthened by the addition of a few lines; and at the poles, it must be proportionally shortened.

I shall now, I think, be able to explain to you the variation of the seasons, and the difference of the length of the days and nights in those seasons; both effects resulting from the same cause.

In moving round the sun, the axis of the earth is not perpendicular to the plane of its orbit. Supposing this round table to represent the plane of the earth's orbit, and this little globe, which has a wire passing through it, representing the axis and poles, we shall call the earth; in moving round the table, the wire is not perpendicular to it, but oblique.

Yes, I understand the earth does not go round
the sun in an upright position, its axis is slanting
or oblique.

MRS. B.

All the lines, which you learnt in your last lesson,
are delineated on this little globe; you must consider the ecliptic as representing the plane of the
earth's orbit; and the equator, which crosses the
ecliptic in two places, shows the degree of obliquity
of the axis of the earth in that orbit, which is exactly 23½ degrees. The points in which the ecliptic
intersects the equator are called nodes.

But I believe I shall make this clearer to you
by revolving the little globe round a candle, which
shall represent the sun. (Plate IX. fig. 2.)

As I now hold it, at A, you see it in the situation in which it is in the midst of summer, or what
is called the summer solstice, which is on the 21st
of June.

EMILY.

You hold the wire awry, I suppose, in order
to show that the axis of the earth is not upright?

MRS. B.

Yes; in summer, the north pole is inclined towards the sun. In this season, therefore, the nor-

thern hemisphere enjoys much more of his rays than the southern. The sun, you see, now shines over the whole of the north frigid zone, and notwithstanding the earth's diurnal revolution, which I imitate by twirling the ball on the wire, it will continue to shine upon it as long as it remains in this situation, whilst the south frigid zone is at the same time completely in obscurity.

### CAROLINE.

That is very strange: I never before heard that there was constant day or night in any part of the world! How much happier the inhabitants of the north frigid zone must be than those of the southern; the first enjoy uninterrupted day, while the last are involved in perpetual darkness.

### MRS. B,

You judge with too much precipitation; examine a little further, and you will find, that the two frigid zones share an equal fate.

We shall now make the earth set off from its position in the summer solstice, and carry it round the sun; observe that the pole is always inclined in the same direction, and points to the same spot in the heavens. There is a fixed star situated near that spot, which is hence called the North Polar star. Now let us stop the earth at B, and examine it in its present situation; it has gone through one quarter of its orbit, and is arrived at that point at which the ecliptic cuts or crosses the equator, and which is called the autumnal equinox.

EMILY.

That is then one of the nodes.

The sun now shines from one pole to the other, just as it would constantly do, if the axis of the earth were perpendicular to its orbit.

MRS. B.

Because the inclination of the axis is now neither towards the sun nor in the contrary direction; at this period of the year, therefore, the days and nights are equal in every part of the earth. But the next step she takes in her orbit, you see, involves the north pole in darkness, whilst it illumines that of the south; this change was gradually preparing as I moved the earth from summer to autumn; the arctic circle, which was at first entirely illumined, began to have short nights, which increased as the earth approached the autumnal equinox; and the instant it passed that point, the long night of the north pole commences, and the south pole begins to enjoy the light of the sun. We shall now make the earth proceed in its orbit, and you may observe that as it advances, the days shorten, and the nights lengthen, throughout the northern hemisphere, until it arrives at the winter solstice, on the 21st of

15

December, when the north frigid zone is entirely in darkness, and the southern has uninterrupted day-light.

### CAROLINE.

Then after all, the sun which I thought so partial, confers his favours equally on all.

### MRS. B.

Not so neither: the inhabitants of the torrid zone have much more heat than we have, as the sun's rays fall perpendicularly on them, while they shine obliquely on the rest of the world, and almost horizontally on the poles; for during their long day of six months, the sun moves round their horizon without either rising or setting; the only observable difference, is that it is more elevated by a few degrees at mid-day, than at mid-night.

### EMILY.

To a person placed in the temperate zone, in the situation in which we are in England, the sun will shine neither so obliquely as it does on the poles, nor so vertically as at the equator; but its rays will fall upon him more obliquely in autumn and winter, than in summer.

### CAROLINE.

And therefore, the inhabitants of the temperate zones, will not have merely one day and one night in the year as happens at the poles, nor will they

K

have equal days and equal nights as at the equator;
but their days and nights will vary in length, at
different times of the year, according as their respective poles incline towards or from the sun, and
the difference will be greater in proportion to their
distance from the equator.

MRS. B.

We shall now follow the earth through the other
half of her orbit, and you will observe, that now
exactly the same effect takes place in the southern
hemisphere, as what we have just remarked in the
northern.   Day commences at the south pole when
night sets in at the north pole; and in every other
part of the southern hemisphere the days are longer
than the nights, while, on the contrary, our nights
are longer than our days.   When the earth arrives
at the vernal equinox, D, where the ecliptic again
cuts the equator, on the 25th of March, she is situated, with respect to the sun, exactly in the same
position, as in the autumnal equinox; and the only
difference with respect to the earth, is, that it is now
autumn in the southern hemisphere, whilst it is
spring with us.

CAROLINE.

Then the days and nights are again every where
equal?

MRS. B.

Yes, for the half of the globe which is enlightened, extends exactly from one pole to the other the day breaks to the north pole, and the sun sets to the south pole; but in every other part of the globe, the day and night is of twelve hours length, hence the word equinox, which is derived from the Latin, meaning equal night.

As the earth proceeds towards summer, the days lengthen in the northern hemisphere, and shorten in the southern, till the earth reaches the summer solstice, when the north frigid zone is entirely illumined, and the southern is in complete darkness; and we have now brought the earth again to the spot from whence we first accompanied her.

#### EMILY.

This is indeed, a most satisfactory explanation of the seasons; and the more I learn, the more I admire the simplicity of means by which such wonderful effects are produced.

#### MRS. B.

I know not which is most worthy of our admiration, the cause, or the effect of the earth's revolution round the sun. The mind can find no object of contemplation, more sublime, than the course of this magnificent globe, impelled by the combined powers of projection and attraction to roll in one invariable course around the source of light and heat: and what can be more delightful than the beneficent effects of this vivifying power on its attendant planet. It is at once the grand principle which animates and fecundates nature.

### EMILY.

There is one circumstance in which this little ivory globe appears to me to differ from the earth; it is not quite dark on that side of it, which is turned from the candle, as is the case with the earth when neither moon nor stars are visible.

### MRS. B.

This is owing to the light of the candle being reflected by the walls of the room on every part of the globe, consequently that side of the globe on which the candle does not directly shine, is not in total darkness. Now the skies have no walls to reflect the sun's light on that side of our earth which is in darkness.

### CAROLINE.

I beg your pardon, Mrs. B., I think that the moon and stars answer the purpose of walls in reflecting the sun's light to us in the night.

### MRS. B.

Very well, Caroline; that is to say, the moon and planets; for the fixed stars, you know shine by their own light.

EMILY.

You say, that the superior heat of the equatorial parts of the earth, arises from the rays falling perpendicularly on those regions, whilst they fall obliquely on these more northern regions; now I do not understand why perpendicular rays should afford more heat than oblique rays.

CAROLINE.

You need only hold your hand perpendicularly over the candle, and then hold it sideways obliquely, to be sensible of the difference.

EMILY.

I do not doubt the fact, but I wish to have it explained.

MRS. B.

You are quite right; if Caroline had not been satisfied with ascertaining the fact, without understanding it, she would not have brought forward the candle as an illustration; the reason why you feel so much more heat if you hold your hand perpendicularly over the candle, than if you hold it sideways, is because a stream of heated vapour constantly ascends from the candle, or any other burning body, which being lighter than the air of the room, does not spread laterally but rises perpendicularly, and this led you to suppose that the rays

K 3

were hotter in the latter direction. Had you reflected, you would have discovered that rays issuing from the candle sideways, are no less perpendicular to your hand when held opposite to them,
than the rays which ascend when your hand is held
over them.

The reason why the sun's rays afford less
heat when in an oblique direction than when
perpendicular, is because fewer of them fall upon
an equal portion of the earth; this will be understood better by referring to Plate X. fig. 1., which
represents two equal portions of the sun's rays,
shining upon different parts of the earth. Here
it is evident, that the same quantity of rays fall on
the space A B, as fall on the space B C; and as
A B is less than B C, the heat and light will be
much stronger in the former than in the latter;
A B, you see, represents the equatorial regions,
where the sun shines perpendicularly; and B C,
the temperate and frozen climates, where his rays
fall more obliquely.

EMILY.

This accounts not only for the greater heat of
the equatorial regions, but for the greater heat of
summer; as the sun shines less obliquely in summer than in winter.

MRS. B.

This you will see exemplified in figure 2, in
which the earth is represented, as it is situated on

PLATE X.

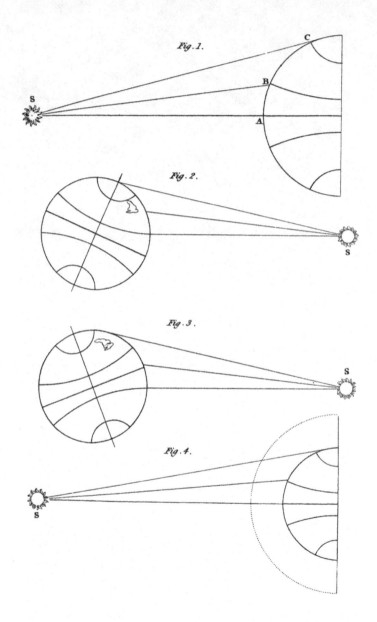

Fig. 1.

Fig. 2.

Fig. 3.

Fig. 4.

Published by Longman & C.º June 1.ˢᵗ 1819.

Lowry Sc.

the 21st of June, and England receives less oblique and consequently a greater number of rays, than at any other season; and figure 3, shows the situation of England on the 21st of December, when the rays of the sun fall most obliquely upon her. But there is also another reason why oblique rays give less heat, than perpendicular rays; which is, that they have a greater portion of the atmosphere to traverse; and though it is true, that the atmosphere is itself a transparent body, freely admitting the passage of the sun's rays, yet it is always loaded more or less with dense and foggy vapour, which the rays of the sun cannot easily penetrate; therefore the greater the quantity of atmosphere the sun's rays have to pass through in their way to the earth, the less heat they will retain when they reach it. This will be better understood, by referring to fig. 4. The dotted line round the earth, describes the extent of the atmosphere, and the lines which proceed from the sun to the earth, the passage of two equal portions of the sun's rays to the equatorial and polar regions; the latter you see, from its greater obliquity passes through a greater extent of atmosphere.

CAROLINE.

And this, no doubt, is the reason why the sun in the morning and the evening gives so much less heat, than at mid-day.

K 4

The diminution of heat, morning and evening, is certainly owing to the greater obliquity of the sun's rays; and as such they are affected by both the causes, which I have just explained to you; the difficulty of passing through a foggy atmosphere is perhaps more particularly applicable to them, as mists and vapours are very prevalent about the time of sunrise and sunset. But the diminished obliquity of the sun's rays, is not the sole cause of the heat of summer; the length of the days greatly conduces to it; for the longer the sun is above the horizon, the more heat he will communicate to the earth.

Both the longest days, and the most perpendicular rays, are on the 21st of June; and yet the greatest heat prevails in July and August.

Those parts of the earth which are once heated, retain the heat for some length of time, and the additional heat they receive, occasions an elevation of temperature, although the days begin to shorten, and the sun's rays to fall more obliquely. For the same reason, we have generally more heat at three o'clock in the afternoon, than at twelve when the sun is on the meridian,

EMILY.

And pray, have the other planets the same vicissitudes of seasons, as the earth?

MRS. B.

Some of them more, some less, according as their axes deviate more or less from the perpendicular to the plane of their orbits. The axis of Jupiter is nearly perpendicular to the plane of his orbit; the axes of Mars and of Saturn are each inclined at angles of about sixty degrees; whilst the axis of Venus is believed to be elevated only fifteen or twenty degrees above her orbit; the vicissitudes of her seasons must therefore be considerably greater than ours. For further particulars respecting the planets, I shall refer you to Bonnycastle's Introduction to Astronomy.

I have but one more observation to make to you relative to the earth's motion, which is, that although we have but 365 days and nights in the year, she performs 366 complete revolutions on her axis during that time.

CAROLINE.

How is that possible? for every complete revolution must bring the same place back to the sun. It is now just twelve o'clock, the sun is, therefore, on our meridian; in twenty-four hours will it not be returned to our meridian again, and will not the earth have made a complete rotation on its axis?

K 5

If the earth had no progressive motion in its orbit whilst it revolves on its axis, this would be the case; but as it advances almost a degree westward in its orbit, in the same time that it completes a revolution eastward on its axis, it must revolve nearly one degree more in order to bring the same meridian back to the sun.

Oh, yes! it will require as much more of a second revolution to bring the same meridian back to the sun, as is equal to the space the earth has advanced in her orbit, that is, nearly a degree; this difference is, however, very little.

These small daily portions of rotation are each equal to the three hundred and sixty-fifth part of a circle, which at the end of the year amounts to one complete rotation.

That is extremely curious. If the earth, then, had no other than its diurnal motion, we should have 366 days in the year.

We should have 366 days in the same period of time that we now have 365; but if we did not revolve round the sun, we should have no natural means of computing years.

You will be surprised to hear, that if time is calculated by the stars instead of the sun, the irregularity which we have just noticed does not occur, and that one complete rotation of the earth on its axis, brings the same meridian back to any fixed star.

EMILY.

That seems quite unaccountable; for the earth advances in her orbit with regard to the fixed stars, the same as with regard to the sun.

MRS. B.

True, but then the distance of the fixed stars is so immense, that our solar system is in comparison to it but a spot, and the whole extent of the earth's orbit but a point; therefore, whether the earth remained stationary, or whether it revolved in its orbit during its rotation on its axis, no sensible difference would be produced with regard to the fixed stars. One complete revolution brings the same meridian back to the same fixed star; hence the fixed stars appear to go round the earth in a shorter time than the sun by three minutes fifty-six seconds of time.

CAROLINE.

These three minutes fifty-six seconds is the time

which the earth takes to perform the additional three
hundred and sixty-fifth part of the circle, in order
to bring the same meridian back to the sun.

MRS. B.

Precisely.  Hence the stars gain every day three
minutes fifty-six seconds on the sun, which makes
them rise that portion of time earlier every day.

When time is calculated by the stars it is called
sidereal time, when by the sun solar or apparent
time.

CAROLINE.

Then a sidereal day is three minutes fifty-six
seconds shorter than a solar day of twenty-four
hours.

MRS. B.

I must also explain to you what is meant by a
sidereal year.

The common year, called the solar or tropical
year, containing 365 days, five hours, forty-eight
minutes, and fifty-two seconds, is measured from
the time the sun sets out from one of the equinoxes,
or solstices, till it returns to the same again; but
this year is completed before the earth has finished
one entire revolution in its orbit.

EMILY.

I thought that the earth performed one complete

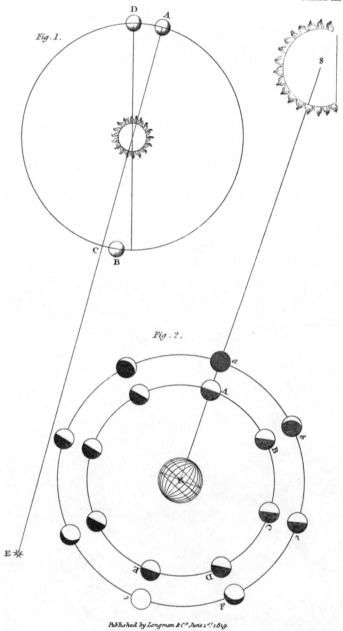

PLATE XI.

Fig. 1.

D   A

C   B

S

Fig. 2.

a

A

b

B

c

C

D

E

Published by Longman & Cᵒ June 1ˢᵗ 1819.

revolution in its orbit every year; what is the reason of this variation?

It is owing to the spheroidal figure of the earth. The elevation about the equator produces much the same effect as if a similar mass of matter, collected in the form of a moon, revolved round the equator. When this moon acted on the earth in conjunction with or in opposition to the sun, variations in the earth's motion would be occasioned, and these variations produce what is called the precession of the equinoxes.

What does that mean? I thought the equinoctial points, or nodes, were fixed points in the heavens, in which the equator cuts the ecliptic.

These points are not quite fixed, but have an 'apparently retrograde motion, that is to say, instead of being every revolution in the same place, they move backwards. Thus, if the vernal equinox is at A, (fig. 1. plate XI.) the autumnal one will be at B instead of C, and the following vernal equinox at D instead of at A, as would be the case if the equinoxes were stationary at opposite points of the earth's orbit.

So that when the earth moves from one equinox
to the other, though it takes half a year to perform
the journey, it has not travelled through half its
orbit.

MRS. B.

And, consequently, when it returns again to the
first equinox, it has not completed the whole of its
orbit.  In order to ascertain when the earth has
performed an entire revolution in its orbit, we
must observe when the sun returns in conjunction
with any fixed star; and this is called a sidereal
year.  Supposing a fixed star situated at E, (fig. I.
plate XI.) the sun would not appear in conjunction
with it till the earth had returned to A, when it
would have completed its orbit.

EMILY.

And how much longer is the sidereal than the
solar year?

MRS. B.

Only twenty minutes; so that the variation of
the equinoctial points is very inconsiderable.  I
have given them a greater extent in the figure in
order to render them sensible.

In regard to time, I must further add, that the
earth's diurnal motion on an inclined axis, together

with its annual revolution in an elliptic orbit, occasions so much complication in its motion, as to produce many irregularities; therefore true equal time cannot be measured by the sun. A clock, which was always perfectly correct, would in some parts of the year be before the sun, and in other parts after it. There are but four periods in which the sun and a perfect clock would agree, which is the 15th of April, the 16th of June, the 23d of August, and the 24th of December.

EMILY.

And is there any considerable difference between solar time and true time?

MRS. B.

The greatest difference amounts to between fifteen and sixteen minutes. Tables of equation are constructed for the purpose of pointing out and correcting these differences between solar time and equal or mean time, which is the denomination given by astronomers to true time.

# CONVERSATION IX.

## ON THE MOON.

OF THE MOON'S MOTION.— PHASES OF THE MOON.
—ECLIPSES OF THE MOON.—ECLIPSES OF JUPI-
TER'S MOONS.—OF THE LATITUDE AND LONGI-
TUDE.— OF THE TRANSITS OF THE INFERIOR
PLANETS.— OF THE TIDES.

MRS. B.

WE shall to-day confine our attention to the
moon, which offers many interesting phenomena.

The moon revolves round the earth in the space
of about twenty-nine days and a half, in an orbit
nearly parallel to that of the earth, and accompanies
us in our revolution round the sun.

EMILY.

Her motion, then, must be rather of a complicated nature; for as the earth is not stationary,
but advances in her orbit whilst the moon goes

round her, the moon must proceed in a sort of progressive circle.

That is true; and there are also other circumstances which interfere with the simplicity and regularity of the moon's motion, but which are too intricate for you to understand at present.

The moon always presents the same face to us, by which it is evident that she turns but once upon her axis, while she performs a revolution round the earth; so that the inhabitants of the moon have but one day and one night in the course of a lunar month.

We afford them, however, the advantage of a magnificent moon to enlighten their long nights.

That advantage is but partial; for since we always see the same hemisphere of the moon, the inhabitants of that hemisphere alone can perceive us.

One half of the moon then enjoys our light every night, while the other half has constantly nights of darkness.  If there are any astronomers in those

regions, they would doubtless be tempted to visit
the other hemisphere, in order to behold so grand
a luminary as we must appear to them. But, pray,
do they see the earth under all the changes which
the moon exhibits to us?

MRS. B.

Exactly so. These changes are called the phases
of the moon, and require some explanation. In
fig. 2. plate XI. let us say that S represents the sun,
E the earth, and A B C D the moon in different
parts of her orbit. When the moon is at A, her
dark side being turned towards the earth, we shall
not see her as at *a;* but her disappearance is of very
short duration, and as she advances in her orbit we
perceive her under the form of a new moon: when
she has gone through one-eighth of her orbit at B,
one quarter of her enlightened hemisphere will be
turned towards the earth, and she will then appear
horned as at *b:* when she has performed one quarter of her orbit, she shows us one half of her enlightened side as at *c;* at *d* she is said to be
gibbous, and at *e* the whole of the enlightened side
appears to us, and the moon is at full. As she
proceeds in her orbit she becomes again gibbous,
and her enlightened hemisphere turns gradually
away from us till she completes her orbit and disappears, and then again resumes her form of a new
moon.

When the moon is at full, or a new moon, she is said to be in conjunction with the sun, as they are then both in the same direction with regard to the earth; when at her quarters she is said to be in opposition to the sun.

**EMILY.**

Are not the eclipses produced by the moon passing between the sun and the earth?

**MRS. B.**

Yes; when the moon passes between the sun and the earth, she intercepts his rays, or in other words, casts a shadow on the earth, then the sun is eclipsed, and the day light gives place to darkness, while the moon's shadow is passing over us.

When, on the contrary, the earth is between the sun and the moon, it is we who intercept the sun's rays, and cast a shadow on the moon; the moon is then darkened, she disappears from our view, and is eclipsed.

**EMILY.**

But as the moon goes round the earth every month, she must be once during that time between the earth and the sun, and the earth must likewise be once between the sun and the moon, and yet we have not a solar and a lunar eclipse every month?

212 ON THE MOON.

MRS. B.

The orbits of the earth and moon are not exactly parallel, but cross or intersect each other; and the moon generally passes either above or below the earth when she is in conjunction with the sun, and does therefore intercept the sun's rays, and produce an eclipse; for this can take place only when the earth and moon are in conjunction in that part of their orbits which cross each other, (called the nodes of their orbits) because it is then only, that they are both in a right line with the sun.

EMILY.

And a partial eclipse takes place, I suppose, when the moon in passing by the earth, is not sufficiently above or below the earth's shadow entirely to escape it?

MRS. B.

Yes, one edge of her disc then dips into the shadow, and is eclipsed; but as the earth is larger than the moon, when the eclipse happens precisely at the nodes, they are not only total, but last for some length of time.

When the sun is eclipsed, the total darkness is confined to one particular part of the earth, evidently showing that the moon is smaller than the earth, since she cannot entirely skreen it from the sun. In fig. 1. pl. XII. you will find a solar eclipse

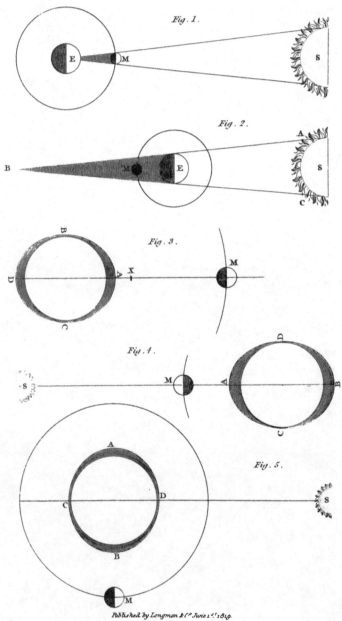

PLATE XII.

Fig. 1.

Fig. 2.

Fig. 3.

Fig. 4.

Fig. 5.

Published by Longman & Co June 1.st 1819.

Lowry Sc.

described; S is the sun, M the moon, and E the earth; and the moon's shadow, you see, is not large enough to cover the earth. The lunar eclipses on the contrary are visible from every part of the earth, where the moon is above the horizon; and we discover by the length of time which the moon is in passing through the earth's shadow, that it would be sufficient to eclipse her totally, were she 47 times her actual size; it follows therefore, that the earth is 47 times the size of the moon.

In fig. 2. S represents the sun, which pours forth rays of light in straight lines in every direction. E is the earth, and M the moon. Now a ray of light coming from one extremity of the sun's disk in the direction A B, will meet another coming from the opposite extremity in the direction C B; the shadow of the earth cannot therefore extend beyond B; as the sun is larger than the earth, the shadow of the latter is conical, or the figure of a sugar loaf; it gradually diminishes, and is much smaller than the earth where the moon passes through it, and yet we find the moon to be not only totally eclipsed, but some length of time in darkness, and hence we are enabled to ascertain its real dimensions.

EMILY.

When the moon eclipses the sun to us, we must be eclipsed to the moon?

16

MRS. B.

Certainly; for if the moon intercepts the sun's rays, and casts a shadow on us, we must necessarily disappear to the moon, but only partially, as in fig. 1.

CAROLINE.

There must be a great number of eclipses in the distant planets, which have so many moons?

MRS. B.

Yes, few days pass without an eclipse taking place: for among the number of satellites, one or other of them are continually passing either between their planet and the sun, or between the planet and each other. Astronomers are so well acquainted with the motion of the planets and their satellites, that they have calculated not only the eclipses of our moon, but those of Jupiter, with such perfect accuracy, that it has afforded a means of ascertaining the longitude.

CAROLINE.

But is it not very easy to find both the latitude and longitude of any place by a map or globe?

MRS. B.

If you know where you are situated, there is no difficulty in ascertaining the latitude or longitude of the place by referring to a map; but supposing

that you had been a length of time at sea, inter-
rupted in your course by storms, a map would afford you very little assistance in discovering where
you were.

CAROLINE.

Under such circumstances, I confess I should be
equally at a loss to discover either latitude or longitude.

MRS. B.

The latitude may be easily found by taking the
altitude of the pole; that is to say, the number of
degrees that it is elevated above the horizon, for
the pole appears more elevated as we approach it,
and less as we recede from it.

CAROLINE.

But unless you can see the pole how can you take
its altitude?

MRS. B.

The north pole points constantly towards one
particular part of the heavens in which a star is situated, called the Polar Star; this star is visible on
clear nights, from every part of the northern hemisphere, the altitude of the polar star, is therefore the
same number of degrees as that of the pole; the
latitude may also be determined by observations

made on the sun or any of the fixed stars; the situation therefore of a vessel at sea, with regard to north and south, is easily ascertained. The difficulty is respecting east and west, that is to say its longitude. As we have no eastern poles from which we can reckon our distance; some particular spot must be fixed upon for that purpose. The English reckon from the meridian of Greenwich, where the royal observatory is situated; in French maps you will find that the longitude is reckoned from Paris.

The rotation of the earth on its axis in 24 hours from west to east occasions, you know, an apparent motion of the sun and stars in the contrary direction, and the sun appears to go round the earth in the space of 24 hours, passing over fifteen degrees or a twenty-fourth part of the earth's circumference every hour; therefore when it is twelve o'clock in London, it is one o'clock in any place situated fifteen degrees to the east of London, as the sun must have passed the meridian of that place an hour before he reaches that of London. For the same reason it is eleven o'clock to any place situated fifteen degrees to the west of London, as the sun will not come to that meridian till an hour later.

If then the captain of a vessel at sea, could know precisely what was the hour at London, he could, by looking at his watch, and comparing it with the hour of the spot in which he was, ascertain the longitude.

EMILY.

But if he had not altered his watch, since he sailed from London, it would indicate the hour it was then in London.

MRS. B.

True; but in order to know the hour of the day of the spot in which he is, the captain of a vessel regulates his watch by the sun when it reaches the meridian.

EMILY.

Then if he had two watches, he might keep one regulated daily, and leave the other unaltered; the former would indicate the hour of the place in which he was situated, and the latter the hour of London; and by comparing them together, he would be able to calculate his longitude.

MRS. B.

You have discovered, Emily, a mode of finding the longitude, which I have the pleasure to tell yuo, is universally adopted: watches of a superior construction, called chronometers, or time-keepers, are used for this purpose; but the best watches are liable to imperfections, and should the time-keeper

L

go too fast, or too slow, there would be no means of ascertaining the error; implicit reliance cannot consequently be placed upon them.

Recourse is therefore had to the eclipses of Jupiter's satellites. A table is made of the precise time at which the several moons are eclipsed to a spectator at London; when they appear eclipsed to a spectator in any other spot, he may, by consulting the table, know what is the hour at London; for the eclipse is visible at the same moment from whatever place on the earth it is seen. He has then only to look at the watch which points out the hour of the place in which he is, and by observing the difference of time there, and at London, he may immediately determine his longitude.

Let us suppose, that a certain moon of Jupiter is always eclipsed at six o'clock in the evening; and that a man at sea consults his watch, and finds that it is ten o'clock at night, where he is situated, at the moment the eclipse takes place; what will be his longitude?

EMILY.

That is four hours later than in London: four times fifteen degrees make 60; he would, therefore, be sixty degrees east of London, for the sun must have passed his meridian before it reaches that of London.

MRS. B.

For this reason the hour is always later than London, when the place is east longitude, and earlier when it is west longitude. Thus the longitude can be ascertained whenever the eclipses of Jupiter's moons are visible.

But it is not only the secondary planets which produce eclipses, for the primary planets near the sun eclipse him to those at a greater distance when they come in conjunction in the nodes of their orbits; but as the primary planets are much longer in performing their course round the sun, than the satellites in going round their primary planets, these eclipses very seldom occur.

Mercury and Venus have however passed in a right line between us and the sun, but being at so great a distance from us, their shadows did not extend so far as the earth; no darkness was therefore produced on any part of our globe; but the planet appeared like a small black spot, passing across the sun's disc; this is called a transit of the planet.

It was by the last transit of Venus, that astronomers were enabled to calculate with some degree of accuracy the distance of the earth from the sun, and the dimensions of the latter.

EMILY.

I have heard that the tides are affected by the

moon, but I cannot conceive what influence it can
have on them.

MRS. B.

They are produced by the moon's attraction,
which draws up the waters in a protuberance.

CAROLINE.

Does attraction act on water more powerfully
than on land? I should have thought it would have
been just the contrary, for land is certainly a more
dense body than water?

MRS. B.

Tides do not arise from water being more
strongly attracted than land, for this certainly is
not the case; but the cohesion of fluids being much
less than that of solid bodies, they more easily yield
to the power of gravity, in consequence of which
the waters immediately below the moon are drawn
up by it in a protuberance, producing a full tide, or
what is commonly called high water, at the spot
where it happens. So far the theory of the tides
is not difficult to understand.

CAROLINE.

On the contrary, nothing can be more simple :
the waters, in order to rise up under the moon,

must draw the waters from the opposite side of the globe, and occasion ebb-tide, or low water in those parts.

### MRS. B.

You draw your conclusion rather too hastily, my dear; for according to your theory, we should have full tide only once in twenty-four hours, that is, every time that we were below the moon, while we find that we have two tides in the course of twenty-four hours, and that it is high-water with us and with our antipodes at the same time.

### CAROLINE.

Yet it must be impossible for the moon to attract the sea in opposite parts of the globe, and in opposite directions at the same time.

### MRS. B.

This opposite tide is rather more difficult to explain, than that which is drawn up beneath the moon; with a little attention, however, I hope I shall be able to make you understand it.

You recollect that the earth and moon are mutually attracted towards a point, their common centre of gravity and of motion; can you tell me what it is that prevents their meeting and uniting at this point?

Their projectile force, which gives them a tendency to fly from this centre.

And is hence called their centrifugal force. Now
we know that the centrifugal force increases in proportion to the distance from the centre of motion.

Yes, I recollect your explaining that to us, and
illustrating it by the motion of the flyers of a windmill, and the spinning of a top.

And it was but the other day you showed us
that bodies weighed less at the equator, than in the
polar regions, in consequence of the increased centrifugal force in the equatorial parts.

Very well. The power of attraction, on the contrary, increases as the distance from the centre of
gravity diminishes; when, therefore, the two centres of gravity and of motion are in the same spot,
as is the case with regard to the moon and the earth,
the centrifugal power and those of attraction, will
be in inverse proportion to each other; that is to

say, where the one is strongest, the other will be weakest.

Those parts of the ocean, then, which are most strongly attracted, will have least centrifugal force; and those parts which are least attracted, will have the greatest centrifugal force.

In order to render the question more simple, let us suppose the earth to be every where covered by the ocean, as represented in (fig. 3. Pl. XII.) M is the moon, A B C D the earth, and X the common centre of gravity and of motion of these two planets. Now the waters on the surface of the earth, about A, being more strongly attracted than any other part, will be elevated; the attraction of the moon at B and C being less, and at D least of all. But the centrifugal force at D, will be greatest, and the waters there, will in consequence have the greatest tendency to recede from the moon; the waters at B and C will have less tendency to recede, and at A least of all. The waters, therefore, at D, will recede furthest, at the same time that they are least attracted, and in consequence will be elevated in a protuberance similar to that at A.

L 4

**EMILY.**

The tide A, then, is produced by the moon's attraction, and increased by the feebleness of the centrifugal power in those parts; and the tide D is produced by the centrifugal force, and increased by the feebleness of the moon's attraction in those parts.

**CAROLINE.**

And when it is high water at A and D, it is low water at B and C: now I think I comprehend the nature of the tides again, though I confess it is not quite so easy as I at first thought.

But, Mrs. B., why does not the sun produce tides as well as the moon; for its attraction is greater than that of the moon?

**MRS. B.**

It would be at an equal distance, but our vicinity to the moon makes her influence more powerful. The sun has, however, a considerable effect on the tides, and increases or diminishes them as it acts in conjunction with, or in opposition to the moon.

**EMILY.**

I do not quite understand that.

**MRS. B.**

The moon is a month in going round the earth;

twice during that time, therefore, at full and at change, she is in the same direction as the sun, both then act in conjunction on the earth, and produce very great tides, called spring tides, as described in fig. 4, at A and B; but when the moon is at the intermediate parts of her orbit, the sun, instead of affording assistance, weakens her power by acting in opposition to it; and smaller tides are produced, called neap tides, as represented in fig. 5.

EMILY.

I have often observed the difference of these tides when I have been at the sea side.

But since attraction is mutual between the moon and the earth, we must produce tides in the moon; and these must be more considerable in proportion as our planet is larger. And yet the moon does not appear of an oval form.

MRS. B.

You must recollect, that in order to render the explanation of the tides clearer, we supposed the whole surface of the earth to be covered with the ocean; but that is not really the case, either with the earth or the moon, and the land which intersects the water destroys the regularity of the effect.

CAROLINE.

True; we may, however, be certain, that when-

L 5

ever it is high water the moon is immediately over
our heads.

MRS. B.

Not so, either; for as a similar effect is produced
on that-part of the globe immediately beneath the
moon, and on that part most distant from it, it cannot be over the heads of the inhabitants of both those
situations at the same time.  Besides, as the orbit
of the moon is very nearly parallel to that of the
earth,  she is never vertical but to the inhabitants of
the torrid zone;  in that climate, therefore, the
tides are greatest, and they diminish as you recede
from it and approach the poles.

CAROLINE.

In the torrid zone, then, I hope you will grant
that the moon is immediately over, or opposite the
spots where it is high water?

MRS. B.

I cannot even admit that; for the ocean naturally
partaking of the earth's motion, in its rotation from
west to east, the moon, in forming a tide, has to
contend against the eastern motion of the waves.
All matter, you know, by its inertia, makes some
resistance to a change of state; the waters, therefore, do not readily yield to the attraction of the
moon, and the effect of her influence is not complete

till three hours after she has passed the meridian, where it is full tide.

EMILY.

Pray what is the reason that the tide is three-quarters of an hour later every day?

MRS. B.

Because it is twenty-four hours and three-quarters before the same meridian on our globe returns beneath the moon. The earth revolves on its axis in about twenty-four hours; if the moon were stationary, therefore, the same part of our globe would, every twenty-four hours, return beneath the moon; but as during our daily revolution the moon advances in her orbit, the earth must make more than a complete rotation in order to bring the same meridian opposite the moon: we are three-quarters of an hour in overtaking her. The tides, therefore, are retarded for the same reason that the moon rises later by three-quarters of an hour every day.

We have now, I think, concluded the observations I had to make to you on the subject of astronomy; at our next interview, I shall attempt to explain to you the elements of hydrostatics.

# CONVERSATION X.

## ON THE MECHANICAL PROPERTIES OF FLUIDS.

DEFINITION OF A FLUID.—DISTINCTION BETWEEN
FLUIDS AND LIQUIDS.—OF NON-ELASTIC FLUIDS.
SCARCELY SUSCEPTIBLE OF COMPRESSION. — OF
THE COHESION OF FLUIDS. — OF THEIR GRAVI-
TATION. — OF THEIR EQULIBRIUM. — OF THEIR
PRESSURE. — OF SPECIFIC GRAVITY. — OF THE
SPECIFIC GRAVITY OF BODIES HEAVIER THAN
WATER. — OF THOSE OF THE SAME WEIGHT AS
WATER. — OF THOSE LIGHTER THAN WATER.—
OF THE SPECIFIC GRAVITY OF FLUIDS.

MRS. B.

WE have hitherto confined our attention to the
mechanical properties of solid bodies, which have
been illustrated, and, I hope, thoroughly impressed
upon your memory, by the conversations we have
subsequently had on astronomy. It will now be

necessary for me to give you some account of the mechanical properties of fluids—a science which is called hydrostatics. A fluid is a substance which yields to the slightest pressure. If you dip your hand into a basin of water, you are scarcely sensible of meeting with any resistance.

EMILY.

The attraction of cohesion is then, I suppose, less powerful in fluids than in solids?

MRS. B.

Yes; fluids, generally speaking, are bodies of less density than solids. From the slight cohesion of the particles of fluids, and the facility with which they slide over each other, it is inferred, that they must be small, smooth, and globular; smooth, because there appears to be little or no friction among them; and globular, because touching each other but by a point would account for the slightness of their cohesion.

CAROLINE.

Pray what is the distinction between a fluid and a liquid?

MRS. B.

Liquids comprehend only one class of fluids.

There is another class distinguished by the name of elastic fluids, or gases, which comprehends the air of the atmosphere, and all the various kinds of air with which you will become acquainted when you study chemistry. Their mechanical properties we shall examine at our next meeting, and confine our attention this morning to those of liquids, or non-elastic fluids.

Water, and liquids in general, are scarcely susceptible of being compressed, or squeezed into a smaller space than that which they naturally occupy. This is supposed to be owing to the extreme minuteness of their particles, which, rather than submit to compression, force their way through the pores of the substance which confines them. This was shown by a celebrated experiment, made at Florence many years ago. A hollow globe of gold was filled with water, and on its being submitted to great pressure, the water was seen to exude through the pores of the gold, which it covered with a fine dew. Fluids gravitate in a more perfect manner than solid bodies; for the strong cohesive attraction of the particles of the latter in some measure counteracts the effect of gravity. In this table, for instance, the cohesion of the particles of wood enables four slender legs to support a considerable weight. Were the cohesion destroyed, or, in other words, the wood converted into a fluid, no support could be afforded by the legs, for the particles no

longer cohering together, each would press separately and independently, and would be brought
to a level with the surface of the earth.

EMILY.

This want of cohesion is then the reason why
fluids can never be formed into figures, or main-
tained in heaps; for though it is true the wind
raises water into waves, they are immediately afterwards destroyed by gravity, and water always finds
its level.

MRS. B.

Do you understand what is meant by the level,
or equilibrium of fluids?

EMILY.

I believe I do, though I feel rather at a loss to
explain it.   Is not a fluid level when its surface is
smooth and flat, as is the case with all fluids when
in a state of rest?

MRS. B.

Smooth, if you please, but not flat; for the definition of the equilibrium of a fluid is, that every
part of the surface is equally distant from the point
to which gravity tends, that is to say, from the centre
of the earth; hence the surface of all fluids must be
bulging, not flat, since they will partake of the

spherical form of the globe. This is very evident in large bodies of water, such as the ocean, but the sphericity of small bodies of water is so trifling, that their surfaces appear flat.

This level, or equilibrium of fluids, is the natural result of their particles gravitating independently of each other; for when any particle of a fluid accidentally finds itself elevated above the rest, it is attracted down to the level of the surface of the fluid, and the readiness with which fluids yield to the slightest impression, will enable the particle by its weight to penetrate the surface of the fluid and mix with it.

CAROLINE.

But I have seen a drop of oil float on the surface of water without mixing with it.

MRS. B.

That is, because oil is a lighter liquid than water. If you were to pour water over it, the oil would rise to the surface, being forced up by the superior gravity of the water. Here is an instrument called a water-level, (fig. 1. plate XIII.) which is constructed upon the principle of the equilibrium of fluids. It consists of a short tube, A B, closed at both ends, and containing a little water; when the tube is not perfectly horizontal the water runs to the lower end, and it is by this

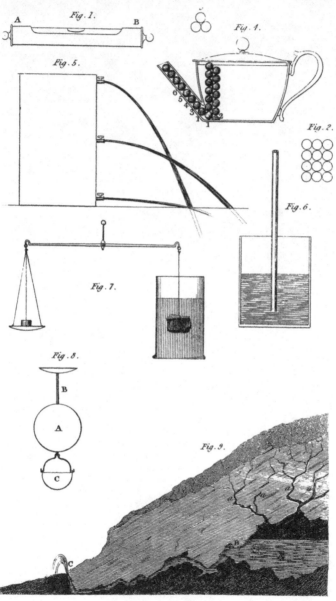

PLATE XIII.

Fig. 1.

A                    B

Fig. 4.

Fig. 5.

Fig. 2.

Fig. 6.

Fig. 7.

Fig. 8.

B

A

C

Fig. 9.

C

Published by Longman & Co. June 1.st 1819.

Lowry Sc.

means that the level of any situation, to which we apply the instrument, is ascertained.

Solid bodies you may, therefore, consider as gravitating in masses, for the strong cohesion of their particles makes them weigh altogether, while every particle of a fluid may be considered as composing a separate mass, gravitating independently of each other.   Hence the resistance of a fluid is considerably less than that of a solid body; for the resistance of the particles acting separately, they are more easily overcome.

### EMILY.

A body of water, in falling, does certainly less injury than a solid body of the same weight.

### MRS. B.

The particles of fluids acting thus independently, press against each other in every direction, not only downwards but upwards, and laterally or sideways; and in consequence of this equality of pressure, every particle remains at rest in the fluid. If you agitate the fluid you disturb this equality of pressure, and the fluid will not rest till its equilibrium is restored.

### CAROLINE.

The pressure downwards is very natural; it is the effect of gravity, one particle weighing upon another presses on it; but the pressure sideways, and particularly the pressure upwards, I cannot understand.

MRS. B.

If there were no lateral pressure, water would not run out of an opening on the side of a vessel. If you fill a vessel with sand, it will not run out of such an opening, because there is scarcely any lateral pressure among its particles.

EMILY.

When water runs out of the side of a vessel, is it not owing to the weight of the water above the opening?

MRS. B.

If the particles of fluids were arranged in regular columns thus, (fig. 2.) there would be no lateral pressure, for when one particle is perpendicularly above the other, it can only press it downwards; but as it must continually happen, that a particle presses between two particles beneath, (fig. 3.) these last must suffer a lateral pressure.

EMILY.

The same as when a wedge is driven into a piece of wood, and separates the parts laterally.

MRS. B.

Yes.   The lateral pressure proceeds, therefore,
entirely from the pressure downwards, or the weight
of the liquid above; and consequently the lower
the orifice is made in the vessel, the greater will be
the velocity of the water rushing out of it.   Here is
a vessel of water (fig. 4.), with three stop cocks at
different heights; we shall open them, and you will
see with what different degrees of velocity the water
issues from them.   Do you understand this,
Caroline?

CAROLINE.

Oh yes.   The water from the upper spout receiving but a slight pressure, on account of its
vicinity to the surface, flows but gently; the second
cock having a greater weight above it, the water is
forced out with greater velocity, whilst the lowest
cock being near the bottom of the vessel, receives
the pressure of almost the whole body of water, and
rushes out with the greatest impetuosity.

MRS. B.

Very well; and you must observe, that as the
lateral pressure is entirely owing to the pressure
downwards, it is not effected by the horizontal dimensions of the vessel, which contains the water,
but merely by its depth; for as every particle acts
independently of the rest, it is only the column of

particles immediately above the orifice that can weigh upon and press out the water.

EMILY.

The breadth and width of the vessel then can be of no consequence in this respect. The lateral pressure on one side, in a cubical vessel, is, I suppose not so great as the pressure downwards.

MRS. B.

No; in a cubical vessel, the pressure downwards will be double the lateral pressure on one side; for every particle at the bottom of the vessel is pressed upon by a column of the whole depth of the fluid, whilst the lateral pressure diminishes from the bottom upwards to the surface, where the particles have no pressure.

CAROLINE.

And from whence proceeds the pressure of fluids upwards? that seems to me the most unaccountable, as it is in direct opposition to gravity.

MRS. B.

And yet it is a consequence of their pressure downwards. When, for example, you pour water into a tea-pot, the water rises in the spout to a level with the water in the pot. The particles of water at the bottom of the pot are pressed upon by

14

the particles above them; to this pressure they will yield, if there is any mode of making way for the superior particles, and as they cannot descend, they will change their direction and rise in the spout.

Suppose the tea-pot to be filled with columns of particles of water similar to that described in fig. 4. the particle 1 at the bottom will be pressed laterally by the particle 2, and by this pressure be forced into the spout, where meeting with the particle 3, it presses it upwards, and this pressure will be continued, from 3 to 4, from 4 to 5, and so on, till the water in the spout has risen to a level with that in the pot.

#### EMILY.

If it were not for this pressure upwards, forcing the water to rise in the spout, the equilibrum of the fluid would be destroyed.

#### CAROLINE.

True; but then a tea-pot is wide and large, and the weight of so great a body of water as the pot will contain, may easily force up and support so small a quantity as will fill the spout. But would the same effect be produced if the spout and the pot were of equal dimensions?

#### MRS. B.

Undoubtedly it would. You may even reverse

the experiment by pouring water into the spout, and you will find that the water will rise in the pot to a level with that in the spout; for the pressure of the small quantity of water in the spout will force up and support the larger quantity in the pot. In the pressure upwards, as well as that laterally, you see that the force of pressure depends entirely on the height, and is quite independent of the horizontal dimensions of the fluid.

As a tea-pot is not transparent, let us try the experiment by filling this large glass goblet by means of this narrow tube, (fig. 6.)

#### CAROLINE.

Look, Emily, as Mrs. B. fills it, how the water rises in the goblet, to maintain an equilibrium with that in the tube.

Now, Mrs. B., will you let me fill the tube by pouring water into the goblet?

#### MRS. B.

That is impossible. However, you may try the experiment, and I doubt not but that you wil be able to account for its failure.

#### CAROLINE.

It is very singular, that if so small a column of water as is contained in the tube can force up and support the whole contents of the goblet; that the weight of all the water in the goblet should not be able to force up the small quantity required to fill the tube: — oh, I see now the reason, the water in the goblet cannot force that in the tube above its level, and as the end of the tube is considerably higher than the goblet, it can never be filled by pouring water into the goblet.

### MRS. B.

And if you continue to pour water into the goblet when it is full, the water will run over instead of rising above the level in the tube.

I shall now explain to you the meaning of the *specific gravity* of bodies.

### CAROLINE.

What! is there another species of gravity with which we are not yet acquainted?

### MRS. B.

No; the specific gravity of a body, means simply its weight compared with that of another body of the same size. When we say, that substances such as lead and stones are heavy, and that others, such as paper and feathers are light, we speak comparatively; that is to say, that the first are heavy, and the latter light, in comparison with the generality of substances in nature. Would you call wood and chalk light or heavy bodies?

CAROLINE.

Some kinds of wood are heavy certainly, as oak and mahogany; others are light, as deal and box.

EMILY.

I think I should call wood in general a heavy body, for deal and box are light only in comparison to wood of a heavier description. I am at a loss to determine whether chalk should be ranked as a heavy or a light body; I should be inclined to say the former, if it was not that it is lighter than most other minerals. I perceive, that we have but vague notions of light and heavy. I wish there was some standard of comparison, to which we could refer the weight of all other bodies.

MRS. B.

The necessity of such a standard has been so much felt, that a body has been fixed upon for this purpose. What substance do you think would be best calculated to answer this end?

CAROLINE.

It must be one generally known and easily obtained, lead or iron, for instance.

MRS. B.

All the metals expand by heat, and condense by cold. A piece of lead, let us say a cubic inch for instance, would have less specific gravity in summer than in winter; for it would be more dense in the latter season.

CAROLINE.

But, Mrs. B., if you compare the weight of equal quantities of different bodies, they will all be alike. You know the old saying, that a pound of feathers is as heavy as a pound of lead?

MRS. B.

When therefore we compare the weight of different kinds of bodies, it would be absurd to take quantities of equal *weight*, we must take quantities of equal *bulk;* pints or quarts, not ounces or pounds.

CAROLINE.

Very true; I perplexed myself by thinking that quantity referred to weight, rather than to measure. It is true, it would be as absurd to compare bodies of the same size in order to ascertain which was largest, as to compare bodies of the same weight in order to discover which was heaviest.

MRS. B.

In estimating the specific gravity of bodies, therefore, we must compare equal bulks, and we shall find that their specific gravity will be proportional

M

to their weights.   The body which has been adopted as a standard of reference is distilled water.

EMILY.

I am surprised that a fluid should have been chosen for this purpose, as it must necessarily be contained in some vessel, and the weight of the vessel will require to be deducted.

MRS. B.

In order to learn the specific gravity of a solid body, it is not necessary to put a certain measure of it in one scale, and an equal measure of water into the other scale; but simply to weigh the body under trial in water.   If you weigh a piece of gold in a glass of water, will not the gold displace just as much water, as is equal to its own bulk?

CAROLINE.

Certainly, where one body is, another cannot be at the same time; so that a sufficient quantity of water must be removed, in order to make way for the gold.

MRS. B.

Yes, a cubic inch of water to make room for a cubic inch of gold; remember that the bulk alone is to be considered, the weight has nothing to do with the quantity of water displaced, for an inch of gold does not occupy more space, and therefore will not displace more water than an inch of ivory, or any other substance, that will sink in water.

Well, you will perhaps be surprised to hear that the gold will weigh less in water, than it did out of it.

### EMILY.

And for what reason?

### MRS. B.

On account of the upward pressure of the par ticles of water, which in some measure supports the gold, and by so doing, diminishes its weight.    If the body immersed in water was of the same weight as that fluid, it would be wholly supported by it, just as the water which it displaces was supported previous to its making way for the solid body.    If the body is heavier than the water, it cannot be wholly supported by it; but the water will offer some resistance to its descent.

### CAROLINE.

And the resistance which water offers to the descent of heavy bodies immersed in it, (since it proceeds from the upward pressure of the particles of the fluid,) must in all cases, I suppose, be the same?

### MRS. B.

Yes; the resistance of the fluid is proportioned to

M 2

the bulk, and not to the weight of the body immersed in it; all bodies of the same size, therefore, lose the same quantity of their weight in water. Can you form any idea what this loss will be?

### EMILY.

I should think it would be equal to the weight of the water displaced; for, since that portion of the water was supported before the immersion of the solid body, an equal weight of the solid body will be supported.

### MRS. B.

You are perfectly right: a body weighed in water loses just as much of its weight, as is equal to that of the water it displaces; so that if you were to put the water displaced into the scale to which the body is suspended, it would restore the balance.

You must observe, that when you weigh a body in water, in order to ascertain its specific gravity, you must not sink the bason of the balance in the water; but either suspend the body to a hook at the bottom of the bason, or else take off the bason, and suspend it to the arm of the balance. (fig. 7.) Now suppose that a cubic inch of gold weighed 19 ounces out of water, and lost one ounce of its weight by being weighed in water, what would be its specific gravity?

CAROLINE.

The cubic inch of water it displaced must weigh that one ounce; and as a cubic inch of gold weighs 19 ounces, gold is 19 times as heavy as water.

EMILY.

I recollect having seen a table of the comparative weights of bodies, in which gold appeared to me to be estimated at 19 thousand times the weight of water.

MRS. B.

You misunderstood the meaning of the table. In the estimation you allude to, the weight of water was reckoned at 1000. You must observe, that the weight of a substance when not compared to that of any other, is perfectly arbitrary; and when water is adopted as a standard, we may denominate its weight by any number we please; but then the weight of all bodies tried by this standard must be signified by proportional numbers.

CAROLINE.

We may call the weight of water, for example, one, and then that of gold would be nineteen; or if we chose to call the weight of water 1000, that of gold would be 19,000. In short, the specific gravity means how much more a body weighs than an equal bulk of water.

M 3

MRS. B.

It is rather the weight of a body compared with that of water; for the specific gravity of many substances is less than that of water.

CAROLINE.

Then you cannot ascertain the specific gravity of such substances in the same manner as that of gold; for a body that is lighter than water will float on its surface without displacing any water.

MRS. B.

If a body were absolutely light, it is true that it would not displace a drop of water, but the bodies we are treating of have all some weight, however small; and will, therefore, displace some quantity of water. If the body be lighter than water, it will not sink to a level with the surface of the water, and therefore it will not displace so much water as is equal to its bulk; but it will displace as much as is equal to its weight. A ship, you must have observed, sinks to some depth in water, and the heavier it is laden the deeper it sinks, as it always displaces a quantity of water equal to its weight.

CAROLINE.

But you said just now, that in the immersion of gold, the bulk, and not the weight of body, was to be considered.

That is the case with all substances which are heavier than water; but since those which are lighter do not displace so much as their own bulk, the quantity they displace is not a test of their specific gravity.

In order to obtain the specific gravity of a body which is lighter than water, you must attach to it a heavy one, whose specific gravity is known, and immerse them together; the specific gravity of the lighter body may then be easily calculated.

EMILY.

But are there not some bodies which have exactly the same specific gravity as water?

MRS. B.

Undoubtedly; and such bodies will remain at rest in whatever situation they are placed in water. Here is a piece of wood which, by being impregnated with a little sand, is rendered precisely of the weight of an equal bulk of water; in whatever part of this vessel of water you place it, you will find that it will remain stationary.

CAROLINE.

I shall first put it at the bottom; from thence, of course, it cannot rise, because it is not lighter than water. Now I shall place it in the middle of the

M 4

vessel; it neither rises nor sinks, because it is neither lighter nor heavier than the water. Now I will lay it on the surface of the water; but there it sinks a little—what is the reason of that, Mrs. B.?

MRS. B.

Since it is not lighter than the water, it cannot float upon its surface; since it is not heavier than water, it cannot sink below its surface: it will sink, therefore, only till the upper surface of both bodies are on a level, so that the piece of wood is just covered with water. If you poured a few drops of water into the vessel, (so gently as not to increase their momentum by giving them velocity) they would mix with the water at the surface, and not sink lower.

CAROLINE.

This must, no doubt, be the reason why, in drawing up a bucket of water out of a well, the bucket feels so much heavier when it rises above the surface of the water in the well; for whilst you raise it in the water, the water within the bucket being of the same specific gravity as the water on the outside, will be wholly supported by the upward pressure of the water beneath the bucket, and consequently very little force will be required to raise it; but as soon as the bucket rises to the surface of the well, you immediately perceive the increase of weight.

EMILY.

And how do you ascertain the specific gravity of fluids?

MRS. B.

By means of an instrument called an hydrometer, which I will shew you. It consists of a thin glass ball A, (fig. 8. Plate XIII.) with a graduated tube B, and the specific gravity of the liquid is estimated by the depth to which the instrument sinks in it. There is a smaller ball, C, attached to the instrument below, which contains a little mercury; but this is merely for the purpose of equipoising the instrument, that it may remain upright in the liquid under trial.

I must now take leave of you; but there remain yet many observations to be made on fluids: we shall, therefore, resume this subject at our next interview.

# CONVERSATION XI.

## OF SPRINGS, FOUNTAINS, &c.

OF THE ASCENT OF VAPOUR AND THE FORM-
ATION OF CLOUDS. — OF THE FORMATION AND
FALL OF RAIN, &c. — OF THE FORMATION OF
SPRINGS. — OF RIVERS AND LAKES. — OF FOUN-
TAINS.

CAROLINE.

THERE is a question I am very desirous of asking
you respecting fluids, Mrs. B., which has often perplexed me. What is the reason that the great
quantity of rain which falls upon the earth and
sinks into it, does not, in the course of time, in-
jure its solidity? The sun and the wind, I know,
dry the surface, but they have no effect on the
interior parts, where there must be a prodigious
accumulation of moisture.

MRS. B.

Do you not know that, in the course of time, all

the water which sinks into the ground rises out of it again?    It is the same water which successively forms seas, rivers, springs, clouds, rain, and sometimes hail, snow, and ice.    If you will take the trouble of following it through these various changes, you will understand why the earth is not yet drowned by the quantity of water which has fallen upon it since its creation; and you will even be convinced, that it does not contain a single drop more water now, than it did at that period.

Let us consider how the clouds were originally formed:    When the first rays of the sun warmed the surface of the earth, the heat, by separating the particles of water, rendered them lighter than the air.    This, you know, is the case with steam or vapour.    What then ensues?

CAROLINE.

When lighter than the air it will naturally rise; and now I recollect your telling us in a preceding lesson, that the heat of the sun transformed the particles of water into vapour, in consequence of which it ascended into the atsmosphere, where it formed clouds.

MRS. B.

We have then already followed water through two of its transformations: from water it becomes vapour, and from vapour clouds.

M 6

EMILY.

But since this watery vapour is lighter than the air, why does it not continue to rise; and why does it unite again to form clouds?

MRS. B.

Because the atmosphere diminishes in density, as it is more distant from the earth. The vapour, therefore, which the sun causes to exhale, not only from seas, rivers, and lakes, but likewise from the moisture on the land, rises till it reaches a region of air of its own specific gravity; and there, you know, it will remain stationary. By the frequent accession of fresh vapour it gradually accumulates, so as to form those large bodies of vapour, which we call clouds; and these, at length, becoming too heavy for the air to support, they fall to the ground.

CAROLINE.

They do fall to the ground, certainly, when it rains; but, according to your theory, I should have imagined, that when the clouds became too heavy for the region of air in which they were situated to support them, they would descend till they reached a stratum of air of their own weight, and not fall to the earth; for as clouds are formed of vapour, they cannot be so heavy as the lowest regions of the atmosphere, otherwise the vapour would not have risen.

#### MRS. B.

If you examine the manner in which the clouds descend, it will obviate this objection. In falling, several of the watery particles come within the sphere of each other's attraction, and unite in the form of a drop of water. The vapour, thus transformed into a shower, is heavier than any part of the atmosphere, and consequently descends to the earth.

#### CAROLINE.

How wonderfully curious!

#### MRS. B.

It is impossible to consider any part of nature attentively without being struck with admiration at the wisdom it displays; and I hope you will never contemplate these wonders without feeling your heart glow with admiration and gratitude towards their bounteous Author. Observe, that if the waters were never drawn out of the earth, all vegetation would be destroyed by the excess of moisture; if, on the other hand, the plants were not nourished and refreshed by occasional showers, the drought would be equally fatal to them. If the clouds constantly remained in a state of vapour, they might, as you remarked, descend into a heavier stratum of the atmosphere, but could never fall to the ground; or were the power of attraction more than sufficient to convert the vapour into drops, it would transform the cloud into a mass of water, which, instead of nourishing, would destroy the produce of the earth.

Water then ascends in the form of vapour, and descends in that of rain, snow, or hail, all of which ultimately become water. Some of this falls into the various bodies of water on the surface of the globe, the remainder upon the land. Of the latter, part reascends in the form of vapour, part is absorbed by the roots of vegetables, and part descends into the bowels of the earth, where it forms springs.

EMILY.

Is rain and spring-water then the same?

MRS. B.

Yes, originally. The only difference between rain and spring water, consists in the foreign particles which the latter meets with and dissolves in its passage through the various soils it traverses.

CAROLINE.

Yet spring water is more pleasant to the taste, appears more transparent, and, I should have supposed, would have been more pure than rain water.

MRS. B.

No; excepting distilled water, rain water is the

most pure we can obtain; and it is its purity which renders it insipid, whilst the various salts and different ingredients, dissolved in spring water, give it a species of flavour, without in any degree affecting its transparency: and the filtration it undergoes through gravel and sand in the bowels of the earth, cleanses it from all foreign matter which it has not the power of dissolving.

When rain falls on the surface of the earth, it continues making its way downwards through the pores and crevices in the ground. When several drops meet in their subterraneous passage, they unite and form a little rivulet: this, in its progress, meets with other rivulets of a similar description, and they pursue their course together in the bowels of the earth, till they are stopped by some substance which they cannot penetrate.

### CAROLINE.

But you said that water could penetrate even the pores of gold, and they cannot meet with a substance more dense?

### MRS. B.

But water penetrates the pores of gold, only when under a strong compressive force, as in the Florentine experiment; now in its passage towards the centre of the earth, it is acted upon by no other power than gravity, which is not sufficient to make

it force its way even through a stratum of clay.
This species of earth, though not remarkably
dense, being of great tenacity, will not admit the
particles of water to pass.  When water encounters
any substance of this nature, therefore, its progress
is stopped, and the pressure of the accumulating
waters forms a bed, or reservoir.  This will be
more clearly explained by fig. 9. Plate XIII. which
represents a section, or the interior of a hill or
mountain.  A, is a body of water such as I have
described, which, when filled up as high as B, (by
the continual accession of waters it receives from
the ducts or rivulets *a, a, a, a,*) finds a passage out
of the cavity, and, impelled by gravity, it runs on,
till it makes its way out of the ground at the side
of the hill, and there forms a spring, C.

CAROLINE.

Gravity impels downwards towards the centre of
the earth; and the spring in this figure runs in an
horizontal direction.

MRS. B.

Not entirely.  There is some declivity from the
reservoir to the spot where the water issues out of
the ground; and gravity you know will bring
bodies down an inclined plane, as well as in a perpendicular direction.

CAROLINE.

But though the spring may descend, on first issuing, it must afterwards rise to reach the surface of the earth; and that is in direct opposition to gravity.

MRS. B.

A spring can never rise above the level of the reservoir whence it issues; it must, therefore, find a passage to some part of the surface of the earth that is lower or nearer the centre than the reservoir. It is true that, in this figure, the spring rises in its passage from B to C occasionally; but this, I think, with a little reflection, you will be able to account for.

EMILY.

Oh, yes; it is owing to the pressure of fluids upwards, and the water rises in the duct upon the same principle as it rises in the spout of a tea-pot; that is to say, in order to preserve an equilibrium with the water in the reservoir. Now I think I understand the nature of springs: the water will flow through a duct, whether ascending or descending, provided it never rises higher than the reservoir.

MRS. B.

Water may thus be conveyed to every part of a

town, and to the upper part of the houses, if it is
originally brought from a height superior to any to
which it is conveyed.    Have you never observed,
when the pavement of the streets has been mending,
the pipes whiwh serve as ducts for the conveyance
of the water through the town?

#### EMILY.

Yes, frequently; and I have remarked that when
any of these pipes have been opened, the water
rushes upwards from them with great velocity,
which, I suppose, proceeds from the pressure of
the water in the reservoir, which forces it out.

#### CAROLINE.

I recollect having once seen a very curious glass,
called Tantalus's cup; it consists of a goblet, containing a small figure of a man, and whatever quantity of water you pour into the goblet, it never rises
higher than the breast of the figure.    Do you know
how that is contrived?

#### MRS. B.

It is by means of a syphon, or bent tube, which
is concealed in the body of the figure.    It rises
through one of the legs as high as the breast, and
there turning descends through the other leg, and
from thence through the foot of the goblet, where
the water runs out. (fig. 1. Plate XIV.)    When

PLATE XIV.

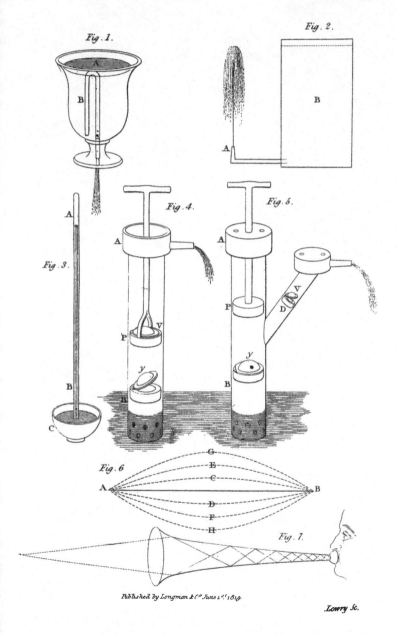

Fig. 1.

Fig. 2.

Fig. 3.

Fig. 4.

Fig. 5.

Fig. 6.

Fig. 7.

Published by Longman & Co. June 1st 1819.

Lowry Sc.

you pour water into the glass A, it must rise in the syphon B, in proportion as it rises in the glass; and when the glass is filled to a level with the upper part of the syphon, the water will run out through the other leg of the figure, and will continue running out, as fast as you pour it in; therefore the glass can never fill any higher.

### EMILY.

I think the new well that has been made at our country-house, must be of that nature. We had a great scarcity of water, and my father has been at considerable expense to dig a well; after penetrating to a great depth before water could be found, a spring was at length discovered, but the water rose only a few feet above the bottom of the well; and sometimes it is quite dry.

### MRS. B.

This has, however, no analogy to Tantalus's cup, but is owing to the very elevated situation of your country-house.

### EMILY.

I believe I guess the reason. There cannot be a reservoir of water near the summit of a hill; as in such a situation, there will not be a sufficient number of rivulets formed to supply one; and without a

reservoir, there can be no spring. In such situations, therefore, it is necessary to dig very deep,
in order to meet with a spring; and when we give
it vent, it can rise only as high as the reservoir
from whence it flows, which will be but little, as the
reservoir must be situated at some considerable
depth below the summit of the hill.

CAROLINE.

Your explanation appears very clear and satisfactory; but I can contradict it from experience. At
the very top of a hill, near our country house, there
is a large pond, and, according to your theory, it
would be impossible there should be springs in such
a situation to supply it with water. Then you
know that I have crossed the Alps, and I can assure
you, that there is a fine lake on the summit of
Mount Cenis, the highest mountain we passed over.

MRS. B.

Were there a lake on the summit of Mount Blanc,
which is the highest of the Alps, it would indeed
be wonderful. But that on Mount Cenis, is not at
all contradictory to our theory of springs; for this
mountain is surrounded by others, much more elevated, and the springs which feed the lake must
descend from reservoirs of water formed in those
mountains. This must also be the case with the

pond on the top of the hill: there is doubtless some more considerable hill in the neighbourhood, which supplies it with water.

### EMILY.

I comprehend perfectly, why the water in our well never rises high; but I do not understand why it should occasionally be dry.

### MRS. B.

Because the reservoir from which it flows, being in an elevated situation, is but scantily supplied with water; after a long drought, therefore, it may be drained, and the spring dry, till the reservoir be replenished by fresh rains. It is not uncommon to see springs flow with great violence in wet weather, and at other times be perfectly dry.

### CAROLINE.

But there is a spring in our grounds which more frequently flows in dry than in wet weather: how is that to be accounted for?

### MRS. B.

The spring probably comes from a reservoir at a great distance, and situated very deep in the ground; it is, therefore, some length of time before the rain reaches the reservoir, and another considerable portion must elapse, whilst the water is making its way from the reservoir to the surface of the earth; so that the dry weather may probably have succeeded the rains before the spring begins to flow, and the reservoir may be exhausted by the time the wet weather sets in again.

### CAROLINE.

I doubt not but this is the case, as the spring is in a very low situation, therefore the reservoir may be at a great distance from it.

### MRS. B.

Springs which do not constantly flow, are called intermitting, and are occasioned by the reservoir being imperfectly supplied. Independently of the situation, this is always the case when the duct or ducts which convey the water into the reservoir are smaller than those which carry it off.

### CAROLINE.

If it runs out faster than it runs in, it will of course sometimes be empty. And do not rivers also derive their source from springs?

### MRS. B.

Yes, they generally take their source in mountainous countries, where springs are most abundant.

CAROLINE.

I understood you that springs were more rare in elevated situations.

MRS. B.

You do not consider that mountainous countries abound equally with high and low situations. Reservoirs of water, which are formed in the bosom of mountains, generally find a vent either on their declivity, or in the valley beneath; while subterraneous reservoirs formed in a plain, can seldom find a passage to the surface of the earth, but remain concealed, unless discovered by digging a well. When a spring once issues at the surface of the earth it continues its course externally, seeking always a lower ground, for it can no longer rise.

EMILY.

Then what is the consequence, if the spring, or I should now rather call it a rivulet, runs into a situation, which is surrounded by higher ground.

MRS. B.

Its course is stopped, the water accumulates, and it forms a pool, pond, or lake, according to the dimensions of the body of water. The Lake of Geneva, in all probability, owes its origin to the Rhone, which passes through it: if, when this river first entered the valley, which now forms the bed of the Lake, it found itself surrounded by higher grounds, its waters would there accumulate, till they rose to a level with that part of the valley, where the Rhone now continues its course beyond the Lake, and from whence it flows through valleys, occasionally forming other small lakes till it reaches the sea.

EMILY.

And are not fountains of the nature of springs?

MRS. B.

Exactly. A fountain is conducted perpendicularly upwards, by the spout or adjutage A, through which it flows; and it will rise nearly as high as the reservoir B, from whence it proceeds. (Plate XIV. fig. 2.)

CAROLINE.

Why not quite as high?

MRS. B.

Because it meets with resistance from the air in its ascent; and its motion is impeded by friction against the spout, where it rushes out.

EMILY.

But if the tube through which the water rises be smooth, can there be any friction? especially with a fluid, whose particles yield to the slightest impression.

MRS. B.

Friction, (as we observed in a former lesson,) may be diminished by polishing, but can never be entirely destroyed; and though fluids are less susceptible of friction than solid bodies, they are still affected by it.   Another reason why a fountain will not rise so high as its reservoir, is, that as all the particles of water spout from the tube with an equal velocity, and as the pressure of the air upon the exterior particles must diminish their velocity, they will in some degree strike against the under parts, and force them sideways, spreading the column into a head, and rendering it both wider and shorter than it otherwise would be.

At our next meeting, we shall examine the mechanical properties of the air, which being an elastic fluid, differs in many respects from liquids.

N

# CONVERSATION XII.

## ON THE MECHANICAL PROPERTIES OF AIR.

OF THE SPRING OR ELASTICITY OF THE AIR. — OF THE WEIGHT OF THE AIR. — EXPERIMENTS WITH THE AIR PUMP. — OF THE BAROMETER. — MODE OF WEIGHING AIR. — SPECIFIC GRAVITY OF AIR. — OF PUMPS. — DESCRIPTION OF THE SUCKING PUMP. — DESCRIPTION OF THE FORCING PUMP.

MRS. B.

AT our last meeting we examined the properties of fluids in general, and more particularly of such fluids as are called liquids.

There is another class of fluids, distinguished by the name of aëriform or elastic fluids, the principal of which is the air we breathe, which surrounds the earth, and is called the atmosphere.

10

EMILY.

There are then other kinds of air, besides the atmosphere?

MRS. B.

Yes, a great variety; but they differ only in their chemical, and not in their mechanical properties; and as it is the latter we are to examine, we shall not at present inquire into their composition, but confine our attention to the mechanical properties of elastic fluids in general.

CAROLINE.

And from whence arises this difference?

MRS. B.

There is no attraction of cohesion between the particles of elastic fluids; so that the expansive power of heat has no adversary to contend with but gravity; any increase of temperature, therefore, expands elastic fluids prodigiously, and a diminution proportionally condenses them.

The most essential point in which air differs from other fluids, is by its spring or elasticity; that is to say, its power of increasing or diminishing in bulk, according as it is more or less compressed: a power of which I have informed you liquids are almost wholly deprived.

### EMILY.

I think I understand the elasticity of the air very well, from what you formerly said of it *; but what perplexes me is, its having gravity; if it is heavy, and we are surrounded by it, why do we not feel its weight?

### CAROLINE.

It must be impossible to be sensible of the weight of such infinitely small particles, as those of which the air is composed: particles which are too small to be seen, must be too light to be felt.

### MRS. B.

You are mistaken, my dear; the air is much heavier than you imagine; it is true, that the particles which compose it are small; but then, reflect on their quantity: the atmosphere extends to about the distance of 45 miles from the earth, and its gravity is such, that a man of middling stature is computed (when the air is heaviest) to sustain the weight of about 14 tons.

### CAROLINE.

Is it possible! I should have thought such a weight would have crushed any one to atoms.

---

* See page 42.

MRS. B.

That would, indeed, be the case, if it were not for the equality of the pressure on every part of the body; but, when thus diffused, we can bear even a much greater weight, without any considerable inconvenience.  In bathing we support the weight and pressure of the water, in addition to that of the atmosphere; but because this pressure is equally distributed over the body, we are scarcely sensible of it; whilst if your shoulders, your head, or any particular part of your frame were loaded with the additional weight of a hundred pounds, you would soon sink under the fatigue.    Besides this, our bodies contain air, the spring of which counterbalances the weight of the external air, and renders us less sensible of its pressure.

CAROLINE.

But if it were possible to relieve me from the weight of the atmosphere, should I not feel more light and agile?

MRS. B.

On the contrary, the air within you meeting with no external pressure to restrain its elasticity, would distend your body, and at length, bursting the parts which confined it, put a period to your existence.

N 3

CAROLINE.

This weight of the atmosphere, then, which I was so apprehensive would crush me, is, in reality, essential to my preservation.

EMILY.

I once saw a person cupped, and was told that the swelling of the part under the cup was produced by taking away from that part the pressure of the atmosphere; but I could not understand how this pressure produced such an effect.

MRS. B.

The air pump affords us the means of making a great variety of interesting experiments on the weight and pressure of the air: some of them you have already seen. Do you not recollect, that in a vacuum produced within the air-pump, substances of various weights fell to the bottom in the same time; why does not this happen in the atmosphere?

CAROLINE.

I remember you told us it was owing to the resistance which light bodies meet with from the air during their fall.

MRS. B.

Or, in other words, to the support which they

received from the air, and which prolonged the time of their fall.   Now, if the air were destitute of weight, how could it support other bodies, or retard their fall?

I shall now show you some other experiments, which illustrate, in a striking manner, both the weight and elasticity of air.   I shall tie a piece of bladder over this glass receiver, which, you will observe, is open both at the top as well as below.

CAROLINE.

Why do you wet the bladder first?

MRS. B.

It expands by wetting, and contracts in drying; it is also more soft and pliable when wet, so that I can make it fit better, and when dry it will be tighter.   We must hold it to the fire in order to dry; but not too near least it should burst by sudden contraction.   Let us now fix it on the air-pump and exhaust the air from underneath it — you will not be alarmed if you hear a noise?

EMILY.

It was as loud as the report of a gun, and the bladder is burst!   Pray explain how the air is concerned in this experiment.

N 4

MRS. B.

It is the effect of the weight of the atmosphere
on the upper surface of the bladder, when I had
taken away the air from the under surface; so that
there was no longer any reaction to counterbalance
the pressure of the atmosphere on the receiver. You
observed how the bladder was pressed inwards by
the weight of the external air, in proportion as I
exhausted the receiver: and before a complete vacuum was formed, the bladder, unable to sustain the
violence of the pressure, burst with the explosion
you have just heard.

I shall now show you an experiment, which
proves the expansion of the air, contained within a
body when it is relieved from the pressure of the external air.   You would not imagine that there was
any air contained within this shrivelled apple, by its
appearance; but take notice of it when placed within a receiver, from which I shall exhaust the air.

CAROLINE.

How strange! it grows quite plump, and looks
like a fresh-gathered apple.

MRS. B.

But as soon as I let the air again into the receiver,
the apple you see returns to its shrivelled state.
When I took away the pressure of the atmosphere,
the air within the apple expanded and swelled it

out; but the instant the atmospherical air was restored, the expansion of the internal air was checked
and repressed, and the apple shrunk to its former
dimensions.

You may make a similar experiment with this
little bladder, which you see is perfectly flaccid, and
appears to contain no air : in this state, I shall tie up
the neck of the bladder, so that whatever air remains
within it may not escape, and then place it under
the receiver.   Now observe, as I exhaust the receiver, how the bladder distends; this proceeds from
the great dilatation of the small quantity of air which
was inclosed within the bladder when I tied it up;
but as soon as I let the air into the receiver, that
which the bladder contains, condenses and shrinks
into its small compass within the folds of the bladder.

### EMILY.

These experiments are extremely amusing, and
they afford clear proofs both of the weight and
elasticity of the air; but I should like to know
exactly how much the air weighs.

### MRS. B.

A column of air reaching to the top of the atmosphere, and whose base is a square inch, weighs
15 lbs. when the air is heaviest; therefore every
square inch of our bodies sustains a weight of
15 lbs.: and if you wish to know the weight of the

whole of the atmosphere, you must reckon how many square inches there are on the surface of the globe, and multiply them by 15.

#### EMILY.

But are there no means of ascertaining the weight of a small quantity of air?

#### MRS. B.

Nothing more easy. I shall exhaust the air from this little bottle by means of the air-pump; and having emptied the bottle of air, or, in other words, produced a vacuum within it, I secure it by turning this screw adapted to its neck: we may now find the exact weight of this bottle, by putting it into one of the scales of a balance. It weighs you see just two ounces; but when I turn the screw, so as to admit the air into the bottle, the scale which contains it preponderates.

#### CAROLINE.

No doubt the bottle filled with air, is heavier than the bottle void of air; and the additional weight required to bring the scales again to a balance, must be exactly that of the air which the bottle now contains.

#### MRS. B.

That weight, you see, is almost two grains. The

dimensions of this bottle are six cubic inches. Six cubic inches of air, therefore, at the temperature of this room, weighs nearly 2 grains.

### CAROLINE.

Why do you observe the temperature of the room, in estimating the weight of the air.

### MRS. B.

Because heat rarefies air, and renders it lighter; therefore the warmer the air is which you weigh, the lighter it will be.

If you should now be desirous of knowing the specific gravity of this air, we need only fill the same bottle with water, and thus obtain the weight of an equal quantity of water — which you see is 1515 grs.; now by comparing the weight of water to that of air, we find it to be in the proportion of about 800 to 1.

I will show you another instance of the weight of the atmosphere, which I think will please you: you know what a barometer is?

### CAROLINE.

It is an instrument which indicates the state of the weather, by means of a tube of quicksilver; but how, I cannot exactly say.

MRS. B.

It is by showing the weight of the atmosphere.
The barometer is an instrument extremely simple
in its construction: in order that you may understand it, I will show you how it is made. I first fill
a glass tube A B, (fig. 3. Plate XIV.) about three
feet in length, and open only at one end, with mercury; then stopping the open end with my finger, I
immerse it in a cup C, containing a little mercury.

EMILY.

Part of the mercury which was in the tube, I observe, runs down into the cup; but why does not
the whole of it subside in the cup, for it is contrary
to the law of the equilibrium of fluids, that the mercury in the tube should not descend to a level with
that in the cup?

MRS. B.

The mercury that has fallen from the tube into
the cup, has left a vacant space in the upper part
of the tube, to which the air cannot gain access;
this space is therefore a perfect vacuum; and consequently the mercury in the tube is relieved from
the pressure of the atmosphere, whilst that in the
cup remains exposed to it.

CAROLINE.

Oh, now I understand it; the pressure of the air

on the mercury in the cup forces it to rise in the tube, where it sustains no pressure.

### EMILY.

Or rather supports the mercury in the tube, and prevents it from falling.

### MRS. B.

That comes to the same thing; for the power that can support mercury in a vacuum, would also make it ascend when it met with a vacuum.

Thus you see, that the equilibrium of the mercury is destroyed only to preserve the general equilibrium of fluids.

### CAROLINE.

But this simple apparatus is, in appearance, very unlike a barometer.

### MRS. B.

It is all that is essential to a barometer. The tube and the cup or vase are fixed on a board, for the convenience of suspending it; the board is graduated for the purpose of ascertaining the height at which the mercury stands in the tube; and the small moveable metal plate serves to show that height with greater accuracy.

278 MECHANICAL PROPERTIES OF AIR.

And at what height will the weight of the atmosphere sustain the mercury?

MRS. B.

About 28 inches, as you will see by this barometer; but it depends upon the weight of the atmosphere, which varies much according to the state of the weather. The greater the pressure of the air on the mercury in the cup, the higher it will ascend in the tube. Now can you tell me whether the air is heavier in wet or dry weather.

CAROLINE.

Without a moment's reflection, the air must be heaviest in wet weather. It is so depressing, and makes one feel so heavy; while in fine weather, I feel as light as a feather, and as brisk as a bee.

MRS. B.

Would it not have been better to have answered with a moment's reflection, Caroline? It would have convinced you, that the air must be heaviest in dry weather, for it is then, that the mercury is found to rise in the tube, and consequently the mercury in the cup must be most pressed by the air: and you know, that we estimate the dryness and fairness of the weather, by the height of the mercury in the barometer.

CAROLINE.

Why then does the air feel so heavy in bad weather.

MRS. B.

Because it is less salubrious when impregnated with damp. The lungs under these circumstances do not play so freely, nor does the blood circulate so well: thus obstructions are frequently occasioned in the smaller vessels, from which arise colds, asthmas, agues, fevers, &c.

EMILY.

Since the atmosphere diminishes in density in the upper regions, is not the air more rare upon a hill than in a plain; and does the barometer indicate this difference?

MRS. B.

Certainly. The hills in this country are not sufficiently elevated to produce any very considerable effect on the barometer; but this instrument is so exact in its indications, that it is used for the purpose of measuring the height of mountains, and of estimating the elevation of balloons.

EMILY.

And is no inconvenience experienced from the thinness of the air in such elevated situations?

MRS. B.

Oh, yes; frequently. It is sometimes oppressive, from being insufficient for respiration; and the expansion which takes place in the more dense air contained within the body is often painful: it occasions distension, and sometimes causes the bursting of the smaller blood-vessels in the nose and ears. Besides, in such situations, you are more exposed both to heat and cold; for though the atmosphere is itself transparent, its lower regions abound with vapours and exhalations from the earth, which float in it, and act in some degree as a covering, which preserves us equally from the intensity of the sun's rays, and from the severity of the cold.

CAROLINE.

Pray, Mrs.B., is not the thermometer constructed on the same principles as the barometer?

MRS. B.

Not at all. The rise and fall of the fluid in the thermometer is occasioned by the expansive power of heat, and the condensation produced by cold: the air has no access to it. An explanation of it would, therefore, be irrelevant to our present subject.

EMILY.

I have been reflecting, that since it is the

14

weight of the atmosphere which supports the mercury in the tube of a barometer, it would support a column of any other fluid in the same manner.

### MRS. B.

Certainly; but as mercury is heavier than all other fluids, it will support a higher column of any other fluid; for two fluids are in equilibrium, when their height varies inversely as their densities. We find the weight of the atmosphere is equal to sustaining a column of water, for instance, of no less than 32 feet above its level.

### CAROLINE.

The weight of the atmosphere is, then, as great as that of a body of water the depth of 32 feet?

### MRS. B.

Precisely; for a column of air of the height of the atmosphere is equal to a column of water of 32 feet, or one of mercury of 28 inches.

The common pump is constructed on this principle. By the act of pumping, the pressure of the atmosphere is taken off the water, which, in consequence, rises.

The body of a pump consists of a large tube or pipe, whose lower end is immersed in the water which it is designed to raise.  A kind of stopper,

called a piston, is fitted to this tube, and is made to
slide up and down it, by means of a metallic rod
fastened to the centre of the piston.

EMILY.

Is it not similar to the syringe, or squirt, with
which you first draw in, and then force out water?

MRS. B.

It is; but you know that we do not wish to
force the water out of the pump, at the same end of
the pipe at which we draw it in.   The intention of
a pump is to raise water from a spring or well; the
pipe is, therefore, placed perpendicularly over the
water which enters it at the lower extremity, and
it issues at a horizontal spout towards the upper
part of the pump.   The pump, therefore, is rather
a more complicated piece of machinery than the
syringe.

Its various parts are delineated in this figure:
(fig. 4. Plate XIV.) A B is the pipe or body of the
pump, P the piston, V a valve, or little door in
the piston, which, opening upwards, admits the
water to rise through it, but prevents its returning,
and Y a similar valve in the body of the pump.

When the pump is in a state of inaction, the
two valves are closed by their own weight; but
when, by drawing down the handle of the pump,
the piston ascends, it raises a column of air which

rested upon it, and produces a vacuum between the piston and the lower valve Y, the air beneath this valve, which is immediately over the surface of the water, consequently expands, and forces its way through it; the water, then, relieved from the pressure of the air, ascends into the pump. A few strokes of the handle totally excludes the air from the body of the pump, and fills it with water, which, having passed through both the valves, runs out at the spout.

CAROLINE.

I understand this perfectly. When the piston is elevated, the air and the water successively rise in the pump; for the same reason as the mercury rises in the barometer.

EMILY.

I thought that water was drawn up into a pump by suction, in the same manner as water may be sucked through a straw.

MRS. B.

It is so, into the body of the pump; for the power of suction is no other than that of producing a vacuum over one part of the liquid, into which vacuum the liquid is forced, by the pressure of the atmosphere on another part. The action of sucking through a straw, consists in drawing in and confining the breath, so as to produce a vacuum in the mouth; in consequence of which, the air within the straw rushes into the mouth, and is followed by the liquid, into which the lower end of the straw is immersed. The principle, you see, is the same; and the only difference consists in the mode of producing a vacuum. In suction, the muscular powers answer the purpose of the piston and valves.

EMILY.

Water cannot, then, be raised by a pump above 32 feet; for the pressure of the atmosphere will not sustain a column of water above that height.

MRS. B.

I beg your pardon. It is true that there must never be so great a distance as 32 feet from the level of the water in the well, to the valve in the piston, otherwise the water would not rise through that valve; but when once the water has passed that opening, it is no longer the pressure of air on the reservoir which makes it ascend; it is raised by lifting it up, as you would raise it in a bucket, of which the piston formed the bottom. This common pump is, therefore, called the sucking, or lifting-pump, as it is constructed on both these principles. There is another sort of pump, called the forcing-pump: it consists of a forcing power added to the sucking part of the pump. This additional power is exactly on the principle of the syringe: by raising the piston you draw the water into the pump, and by descending it you force the water out.

CAROLINE.

But the water must be forced out at the upper part of the pump; and I cannot conceive how that can be done by descending the piston.

MRS. B.

Figure 5. Pl. XIV. will explain the difficulty. The large pipe A B represents the sucking part of the pump, which differs from the lifting-pump only in its piston P being unfurnished with a valve, in consequence of which the water cannot rise above it. When, therefore, the piston descends, it shuts the valve Y, and forces the water (which has no other vent) into the pipe D: this is likewise furnished with a valve V, which, opening outwards, admits the water, but prevents its return.

The water is thus first raised in the pump, and then forced into the pipe, by the alternate ascending and descending motion of the piston, after a few strokes of the handle to fill the pipe, from whence the water issues at the spout.

It is now time to conclude our lesson. When next we meet, I shall give you some account of wind, and of sound, which will terminate our observations on elastic fluids.

## CAROLINE.

And I shall run into the garden, to have the pleasure of pumping, now that I understand the construction of a pump.

## MRS. B.

And, to-morrow, I hope you will be able to tell me, whether it is a forcing or a common lifting pump.

# CONVERSATION XIII.

## ON WIND AND SOUND.

OF WIND IN GENERAL. — OF THE TRADE WIND. —
OF THE PERIODICAL TRADE WINDS. — OF THE
AËRIAL TIDES. — OF SOUND IN GENERAL. — OF
SONOROUS BODIES. — OF MUSICAL SOUNDS. — OF
CONCORD OR HARMONY, AND MELODY.

MRS. B.

WELL, Caroline, have you ascertained what kind
of pump you have in your garden?

CAROLINE.

I think it must be merely a lifting-pump, because no more force is required to raise the handle
than is necessary to lift its weight; and in a forcing-
pump, by raising the handle, you force the water

into the smaller pipe, and the resistance the water offers must require an exertion of strength to overcome it.

MRS. B.

I make no doubt you are right; for lifting pumps, being simple in their construction, are by far the most common.

I have promised to day to give you some account of the nature of wind.  Wind is nothing more than the motion of a stream or current of air, generally produced by a partial change of temperature in the atmosphere; for when any one part is more heated than the rest, that part is rarefied; the equilibrium is destroyed, and the air in consequence rises.  When this happens, there necessarily follows a motion of the surrounding air towards that part, in order to restore it; this spot, therefore, receives winds from every quarter.  Those who live to the north of it experience a north wind; those to the south, a south wind: — do you comprehend this?

CAROLINE.

Perfectly.  But what sort of weather must those people have, who live on the spot where these winds meet and interfere?

### MRS. B.

They have turbulent and boisterous weather, whirlwinds, hurricanes, rain, lightning, thunder, &c. This stormy weather occurs most frequently in the torrid zone, where the heat is greatest: the air being more rarefied there, than in any other part of the globe is lighter, and consequently ascends; whilst the air about the polar regions is continually flowing from the poles, to restore the equilibrium.

### CAROLINE.

This motion of the air would produce a regular and constant north wind to the inhabitants of the northern hemisphere; and a south wind to those of the southern hemisphere, and continual storms at the equator, where these two adverse winds would meet.

### MRS. B.

These winds do not meet, for they each change their direction before they reach the equator. The sun, in moving over the equatorial regions from east to west, rarefies the air as it passes, and causes the denser eastern air to flow westwards, in order to restore the equilibrium; thus producing a regular east wind about the equator.

CAROLINE.

The air from the west, then, constantly goes to
meet the sun, and repair the disturbance which his
beams have produced in the equilibrium of the atmosphere. But I wonder how you will reconcile
these various winds, Mrs. B.: you first led me to
suppose there was a constant struggle between opposite winds at the equator, producing storm and
tempest; but now I hear of one regular invariable
wind, which must naturally be attended by calm
weather.

EMILY.

I think I comprehend it: do not these winds
from the north and south combine with the easterly
wind about the equator, and form what are called
the trade-winds?

MRS. B.

Just so, my dear. The composition of the two
winds north and east, produces a constant north-
east wind; and that of the two winds south and
east, produces a regular south-east wind: these
winds extend to about thirty degrees on each side
of the equator, the regions further distant from it
experiencing only their respective north and south
winds.

CAROLINE.

But, Mrs. B., if the air is constantly flowing

from the poles to the torrid zone, there must be a deficiency of air in the polar regions?

MRS. B.

The light air about the equator, which expands and rises into the upper regions of the atmosphere, ultimately flows from thence back to the poles, to restore the equilibrium: if it were not for this resource, the polar atmospheric regions would soon be exhausted by the stream of air, which, in the lower strata of the atmosphere, they are constantly sending towards the equator.

CAROLINE.

There is then a sort of circulation of air in the atmosphere; the air in the lower strata flowing from the poles towards the equator, and in the upper strata, flowing back from the equator towards the poles.

MRS. B.

Exactly: I can show you an example of this circulation on a small scale. The air of this room being more rarefied than the external air, a wind or current of air is pouring in from the crevices of the windows and doors, to restore the equilibrium; but the light air with which the room is filled must find some vent, in order to make way for the heavy air which enters. If you set the door a-jar, and hold a candle near the upper part of it, you will find that the flame will be blown outwards, showing that there is a current of air flowing out from the upper part of the room.—Now place the candle on the floor close by the door, and you will perceive, by the inclination of the flame, that there is also a current of air setting into the room.

### CAROLINE.

It is just so; the upper current is the warm light air, which is driven out to make way for the stream of cold dense air which enters the room lower down.

### EMILY.

I have heard, Mrs. B., that the periodical winds are not so regular on land as at sea: what is the reason of that?

### MRS. B.

The land reflects into the atmosphere a much greater quantity of the sun's rays than the water; therefore, that part of the atmosphere which is over the land, is more heated and rarefied than that which is over the sea: this occasions the wind to set in upon the land, as we find that it regularly does on the coast of Guinea, and other countries in the torrid zone.

EMILY.

I have heard much of the violent tempests occasioned by the breaking up of the monsoons; are not they also regular trade-winds?

MRS. B.

They are called periodical trade-winds, as they change their course every half-year. This variation is produced by the earth's annual course round the sun, when the north pole is inclined towards that luminary one half of the year, the south pole the other half. During the summer of the northern hemisphere, the countries of Arabia, Persia, India and China, are much heated, and reflect great quantities of the sun's rays into the atmosphere, by which it becomes extremely rarefied, and the equilibrium consequently destroyed. In order to restore it, the air from the equatorial southern regions, where it is colder, (as well as from the colder northern parts,) must necessarily have a motion towards those parts. The current of air from the equatorial regions produces the trade-winds for the first six months, in all the seas between the heated continent of Asia, and the equator. The other six months, when it is summer in the southern hemisphere, the ocean and countries towards the southern tropic are most heated, and the air over those parts most rarefied: then the air about the equator alters its course, and flows exactly in an opposite direction.

CAROLINE.

This explanation of the monsoons is very curious; but what does their breaking up mean?

MRS. B.

It is the name given by sailors to the shifting of the periodical winds; they do not change their course suddenly, but by degrees, as the sun moves from one hemisphere to the other: this change is usually attended by storms and hurricanes, very dangerous for shipping; so that those seas are seldom navigated at the season of the equinox.

EMILY.

I think I understand the winds in the torrid zone perfectly well; but what is it that occasions the great variety of winds which occur in the temperate zones? for, according to your theory, there should be only north and south winds in those climates.

MRS. B.

Since so large a portion of the atmosphere as is over the torrid zone is in continued agitation, these agitations in an elastic fluid, which yields to the slightest impression, must extend every way to a great distance; the air, therefore, in all climates, will suffer more or less perturbation, according to the situation of the country, the position of mountains, valleys, and a variety of other causes: hence it is easy to conceive, that almost every climate must be liable to variable winds.

On the sea-shore, there is almost always a gentle sea-breeze setting in on the land on a summer's evening, to restore the equilibrium which had been disturbed by reflections from the heated surface of the shore during the day; and when night has cooled the land, and condensed the air, we generally find it towards morning, flowing back towards the sea.

CAROLINE.

I have observed, that the wind, which-ever way it blows, almost always falls about sun-set?

MRS. B.

Because the rarefaction of air in the particular spot which produces the wind, diminishes as the sun declines, and consequently the velocity of the wind abates.

EMILY.

Since the air is a gravitating fluid, is it not af-

o 4

fected by the attraction of the moon and the sun, in the same manner as the waters?

### MRS. B.

Undoubtedly; but the aërial tides are as much greater than those of water, as the density of water exceeds that of air, which, as you may recollect, we found to be about 800 to 1.

### CAROLINE.

What a prodigious protuberance that must occasion! How much the weight of such a column of air must raise the mercury in the barometer!

### EMILY.

As this enormous tide of air is drawn up and supported, as it were, by the moon, its weight and pressure, I should suppose, would be rather diminished than increased?

### MRS. B.

The weight of the atmosphere is neither increased nor diminished by the aërial tides. The moon's attraction augments the bulk as much as it diminishes the weight of the column of air; these effects, therefore, counterbalancing each other, the aërial tides do not affect the barometer.

CAROLINE.

I do not quite understand that.

MRS. B.

Let us suppose that the additional bulk of air at high tide raises the barometer one inch; and on the other hand, that the support which the moon's attraction affords the air diminishes its weight or pressure, so as to occasion the mercury to fall one inch; under these circumstances the mercury must remain stationary. Thus you see, that we can never be sensible of aërial tides by the barometer, on account of the equality of pressure of the atmosphere, whatever be its height.

The existence of aërial tides is not, however, hypothetical; it is proved by the effect they produce on the apparent position of the heavenly bodies; but this I cannot explain to you, till you understand the properties of light.

EMILY.

And when shall we learn them?

MRS. B.

I shall first explain to you the nature of sound, which is intimately connected with that of air; and I think at our next meeting we may enter upon the subject of optics.

We have now considered the effects produced by the wide and extended agitation of the air; but there is another kind of agitation of which the air is susceptible — a sort of vibratory trembling motion, which, striking on the drum of the ear, produces *sound.*

### CAROLINE.

Is not sound produced by solid bodies? The voice of animals, the ringing of bells, musical instruments, are all solid bodies. I know of no sound but that of the wind which is produced by the air.

### MRS. B.

Sound, I assure you, results from a tremulous motion of the air; and the sonorous bodies you enumerate, are merely the instruments by which that peculiar species of motion is communicated to the air.

### CAROLINE.

What! when I ring this little bell, is it the air that sounds, and not the bell?

### MRS. B.

Both the bell and the air are concerned in the production of sound. But sound, strictly speaking, is a perception excited in the mind by the motion of the air on the nerves of the ear; the air, therefore, as

12

well as the sonorous bodies which put it in motion, is only the cause of sound, the immediate effect is produced by the sense of hearing; for without this sense, there would be no sound.

EMILY.

I can with difficulty conceive that. A person born deaf, it is true, has no idea of sound, because he hears none; yet that does not prevent the real existence of sound, as all those who are not deaf can testify.

MRS. B.

I do not doubt the existence of sound to all those who possess the sense of hearing; but it exists neither in the sonorous body nor in the air, but in the mind of the person whose ear is struck by the vibratory motion of the air, produced by a sonorous body.

To convince you that sound does not exist in sonorous bodies, but that air or some other vehicle is necessary to its production, endeavour to ring the little bell, after I have suspended it under a receiver in the air-pump, from which I shall exhaust the air........

CAROLINE.

This is indeed very strange: though I agitate

o 6

it so violently, it does not produce the least
sound.

By exhausting the receiver, I have cut off the
communication between the air and the bell; the
latter, therefore, cannot impart its motion to the air.

Are you sure that it is not the glass, which covers the bell, that prevents our hearing it?

That you may easily ascertain by letting the air
into the receiver, and then ringing the bell.

Very true: I can hear it now almost as loud as
if the glass did not cover it; and I can no longer
doubt but that air is necessary to the production of
sound.

Not absolutely necessary, though by far the most
common vehicle of sound. Liquids, as well as air, are
capable of conveying the vibratory motion of a sonorous body to the organ of hearing; as sound can

be heard under water. Solid bodies also convey sound, as I can soon convince you by a very simple experiment. I shall fasten this string by the middle round the poker; now raise the poker from the ground by the two ends of the string, and hold one to each of your ears: — I shall now strike the poker with a key, and you will find that the sound is conveyed to the ear by means of the strings, in a much more perfect manner than if it had no other vehicle than the air.

That it is, certainly, for I am almost stunned by the noise. But what is a sonorous body, Mrs. B.? for all bodies are capable of producing some kind of sound by the motion they communicate to the air.

Those bodies are called sonorous, which produce clear, distinct, regular, and durable sounds, such as a bell, a drum, musical strings, wind-instruments, &c. They owe this property to their elasticity; for an elastic body, after having been struck, not only returns to its former situation, but having acquired momentum by its velocity, like the pendulum, it springs out on the opposite side. If I draw the string A B, which is made fast at both

ends to C, it will not only return to its original
position, but proceed onwards to D. This is its first
vibration, at the end of which it will retain sufficient
velocity to bring it to E, and back again to F,
which constitutes its second vibration; the third
vibration will carry it only to G and H, and
so on, till the resistance of the air destroys its
motion.

The vibration of a sonorous body gives a tremulous motion to the air around it, very similar to
the motion communicated to smooth water when a
stone is thrown into it. This first produces a
small circular wave around the spot in which the
stone falls; the wave spreads, and gradually communicates its motion to the adjacent waters, producing similar waves to a considerable extent. The
same kind of waves are produced in the air by
the motion of a sonorous body, but with this difference, that as air is an elastic fluid, the motion
does not consist of regularly extending waves, but
of vibrations, and are composed of a motion forwards and backwards, similar to those of the sonorous body.    They differ also in the one taking
place in a plane, the other in all directions.   The
aërial undulations being spherical.

#### EMILY.

But if the air moves backwards as well as forwards, how can its motion extend so as to convey sound to a distance?

The first sphere of undulations which are produced immediately around the sonorous body, by pressing against the contiguous air, condenses it. The condensed air, though impelled forward by the pressure, re-acts on the first set of undulations, driving them back again. The second set of undulations which have been put in motion, in their turn communicate their motion, and are themselves driven back by re-action. Thus there is a succession of waves in the air, corresponding with the succession of waves in the water.

The vibrations of sound must extend much further than the circular waves in water, since sound is conveyed to a great distance.

The air is a fluid so much less dense than water, that motion is more easily communicated to it. The report of a cannon produces vibrations of the air, which extend to several miles around.

Distant sound takes some time to reach us,

since it is produced at the moment the cannon is fired; and we see the light of the flash long before we hear the report.

MRS. B.

The air is immediately put in motion by the firing of a cannon; but it requires time for the vibrations to extend to any distant spot. The velocity of sound is computed to be at the rate of 1142 feet in a second.

CAROLINE.

With what astonishing rapidity the vibrations must be communicated! But the velocity of sound varies, I suppose, with that of the air which conveys it. If the wind sets towards us from the cannon, we must hear the report sooner than if it set the other way.

MRS. B.

The direction of the wind makes less difference in the velocity of sound than you would imagine. If the wind sets from us, it bears most of the aërial waves away, and renders the sound fainter; but it is not very considerably longer in reaching the ear than if the wind blew towards us. This uniform velocity of sound enables us to determine the distance of the object from which it proceeds; as that of a vessel at sea firing a cannon, or that of a thun-

15

der cloud. If we do not hear the thunder till half a minute after we see the lightning, we conclude the cloud to be at the distance of six miles and a half.

### EMILY.

Pray how is the sound of an echo produced?

### MRS. B.

When the aërial vibrations meet with an obstacle, having a hard and regular surface, such as a wall, or rock, they are reflected back to the ear, and produce the same sound a second time; but the sound will then appear to proceed from the object by which it is reflected. If the vibrations fall perpendicularly on the obstacle, they are reflected back in the same line; if obliquely, the sound returns obliquely in the opposite direction, the angle of reflection being equal to the angle of incidence.

### CAROLINE.

Oh, then, Emily, I now understand why the echo of my voice behind our house is heard so much plainer by you than it is by me, when we stand at the opposite ends of the gravel walk. My voice, or rather, I should say, the vibrations of air it occasions, fall obliquely on the wall of the house, and are reflected by it to the opposite end of the gravel walk.

### EMILY.

Very true; and we have observed, that when we stand in the middle of the walk, opposite the house, the echo returns to the person who spoke.

### MRS. B.

Speaking-trumpets are constructed on the principle of the reflection of sound. The voice, instead of being diffused in the open air, is confined within the trumpet; and the vibrations, which spread and fall against the sides of the instrument, are reflected according to the angle of incidence, and fall into the direction of the vibrations which proceed straight forwards. The whole of the vibrations are thus collected into a focus; and if the ear be situated in or near that spot, the sound is prodigiously increased. Figure 7. Plate XIV. will give you a clearer idea of the speaking-trumpet: the reflected rays are distinguished from those of incidence, by being dotted; and they are brought to a focus at F. The trumpet used by deaf persons acts on the same principle; but as the voice enters the trumpet at the large, instead of the small end of the instrument, it is not so much confined, nor the sound so much increased.

### EMILY.

Are the trumpets used as musical instruments also constructed on this principle?

So far as their form tends to increase the sound, they are; but, as a musical instrument, the trumpet becomes itself the sonorous body, which is made to vibrate by blowing into it, and communicates its vibrations to the air.

I will attempt to give you, in a few words, some notion of the nature of musical sounds, which, as you are fond of music, must be interesting to you.

If a sonorous body be struck in such a manner, that its vibrations are all performed in regular times, the vibrations of the air will correspond with them; and striking in the same regular manner on the drum of the ear, will produce the same uniform sensation on the auditory nerve, and excite the same uniform idea in the mind; or, in other words, we shall hear one musical tone.

But if the vibrations of the sonorous body are irregular, there will necessarily follow a confusion of aërial vibrations; for a second vibration may commence before the first is finished, meet it half way on its return, interrupt it in its course, and produce harsh jarring sounds, which are called *discords.*

EMILY.

But each set of these irregular vibrations, if repeated at equal intervals, would, I suppose, produce a musical tone? It is only their irregular succession which makes them interfere, and occasions discord.

### MRS. B.

Certainly. The quicker a sonorous body vibrates, the more acute, or sharp, is the sound produced.

### CAROLINE.

But if I strike any one note of the piano-forte repeatedly, whether quickly or slowly, it always gives the same tone.

### MRS. B.

Because the vibrations of the same string, at the same degree of tension, are always of a similar duration. The quickness or slowness of the vibrations relate to the single tones, not to the various sounds which they may compose by succeeding each other. Striking the note in quick succession, produces a more frequent repetition of the tone, but does not increase the velocity of the vibrations of the string. The duration of the vibrations of strings or chords, depends upon their length, their thickness or weight, and their degree of tension: thus, you find, the low bass notes are produced by long, thick, loose strings; and the high treble notes by short, small, and tight strings.

CAROLINE.

Then the different length and size of the strings of musical instruments, serves to vary the duration of the vibrations, and, consequently, the acuteness or gravity of the notes?

MRS. B.

Yes. Among the variety of tones, there are some which, sounded together, please the ear, producing what we call harmony, or concord. This arises from the agreement of the vibrations of the two sonorous bodies; so that some of the vibrations of each strike upon the ear at the same time. Thus, if the vibrations of two strings are performed in equal times, the same tone is produced by both, and they are said to be in unison.

EMILY.

Now, then, I understand why, when I tune my harp in unison with the piano-forte, I draw the strings tighter if it is too low, or loosen them if it is at too high a pitch: it is in order to bring them to vibrate, in equal times, with the strings of the piano-forte.

MRS. B.

But concord, you know, is not confined to unison; for two different tones harmonize in a variety of cases. If the vibrations of one string (or sonorous body whatever) vibrate in double the time of another, the second vibration of the latter will strike upon the ear at the same instant as the first vibration of the former; and this is the concord of an octave.

If the vibrations of two strings are as two to three, the second vibration of the first corresponds with the third vibration of the latter, producing the harmony called a fifth.

### CAROLINE.

So, then, when I strike the key-note with its fifth, I hear every second vibration of one, and every third of the other at the same time?

### MRS. B.

Yes; and the key-note struck with the fourth is likewise a concord, because the vibrations are as three to four. The vibrations of a major third with the key-note, are as four to five; and those of a minor third, as five to six.

There are other tones which, though they cannot be struck together without producing discord, if struck successively, give us the pleasure which is called melody. Upon these general principles the science of music is founded; but I am not sufficiently acquainted with it to enter any further into it.

We shall now, therefore, take leave of the subject of sound; and, at our next interview, enter upon that of optics, in which we shall consider the nature of vision, light, and colours.

# CONVERSATION XIV.

## ON OPTICS.

OF LUMINOUS, TRANSPARENT, AND OPAQUE BO-
DIES. — OF THE RADIATION OF LIGHT. — OF
SHADOWS. — OF THE REFLECTION OF LIGHT. —
OPAQUE BODIES SEEN ONLY BY REFLECTED
LIGHT. — VISION EXPLAINED. — CAMERA OB-
SCURA. — IMAGE OF OBJECTS ON THE RETINA.

#### CAROLINE.

I LONG to begin our lesson to day, Mrs. B., for I expect that it will be very entertaining.

#### MRS. B.

Optics is certainly one of the most interesting branches of Natural Philosophy, but not one of the easiest to understand; I must therefore beg, that you will give me the whole of your attention.

I shall first inquire, whether you comprehend

the meaning of a *luminous body*, an *opaque body*, and a *transparent body*.

A luminous body is one that shines; an opaque . . . .

Do not proceed to the second, until we have agreed upon the definition of the first. All bodies that shine are not luminous; for a luminous body is one that shines by its own light, as the sun, the fire, a candle, &c.

Polished metal then, when it shines with so much brilliancy, is not a luminous body?

No, for it would be dark if it did not receive light from a luminous body; it belongs, therefore, to the class of opaque or dark bodies, which comprehend all such as are neither luminous nor will admit the light to pass through them.

And transparent bodies, are those which admit the light to pass through them; such as glass and water?

P

You are right. Transparent or pellucid bodies, are frequently called mediums; and the rays of light which pass through them, are said to be transmitted by them.

Light, when emanated from the sun, or any other luminous body, is projected forwards in straight lines in every possible direction: so that the luminous body is not only the general centre from whence all the rays proceed; but every point of it may be considered as a centre which radiates light in every direction. (Fig. 1. Plate XV.)

EMILY.

But do not the rays which are projected in different directions, and cross each other, interfere and impede each other's course?

MRS. B.

Not at all. The particles of light are so extremely minute, that they are never known to interfere with each other. A ray of light, is a single line of light projected from a luminous body; and a pencil of rays, is a collection of rays, proceeding from any one point of a luminous body, as fig. 2.

CAROLINE.

Is light then a substance composed of particles like other bodies?

PLATE XV

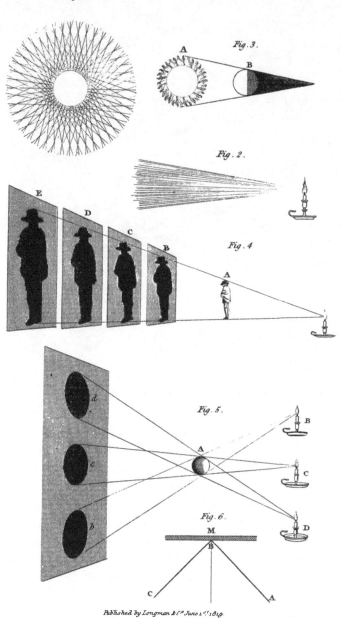

*Fig. 1.*

*Fig. 3.*

*Fig. 2.*

*Fig. 4*

*Fig. 5.*

*Fig. 6.*

Published by Longman & C.º June 1.ˢᵗ 1819.

*Lowry Sc.*

That is a disputed point, upon which I cannot pretend to decide. In some respects, light is obedient to the laws which govern bodies; in others, it appears to be independent of them: thus though its course is guided by the laws of motion, it does not seem to be influenced by those of gravity. It has never been discovered to have weight, though a variety of interesting experiments have been made with a view of ascertaining that point; but we are so ignorant of the intimate nature of light, that an attempt to investigate it would lead us into a labyrinth of perplexity, if not of error; we shall therefore confine our attention to those properties of light which are well ascertained.

Let us return to the examination of the effects of the radiation of light from a luminous body. Since the rays of light are projected in straight lines, when they meet with an opaque body through which they are unable to pass, they are stopped short in their course; for they cannot move in a curve line round the body.

No, certainly; for it would require some other force besides that of projection, to produce motion in a curve line.

The interruption of the rays of light, by the
opaque body, produces, therefore, darkness on the
opposite side of it; and if this darkness fall upon a
wall, a sheet of paper, or any object whatever, it
forms a shadow.

A shadow then, is nothing more than darkness
produced by the intervention of an opaque body,
which prevents the rays of light from reaching an
object behind the opaque body.

Why then are shadows of different degrees of
darkness; for I should have supposed from your
definition of a shadow, that it would have been perfectly black?

It frequently happens that a shadow is produced
by an opaque body interrupting the course of the
rays from one luminous body, while light from
another reaches the space where the shadow is
formed, in which case the shadow is proportionally
fainter. This happens if the opaque body be lighted
by two candles: if you extinguish one of them,
the shadow will be both deeper and more distinct.

CAROLINE.

But yet it will not be perfectly dark.

MRS. B.

Because it is still slightly illumined by light reflected from the walls of the room, and other surrounding objects.

You must observe, also, that when a shadow is
produced by the interruption of rays from a single
luminous body, the darkness is proportional to the
intensity of the light.

EMILY.

I should have supposed the contrary; for as the
light reflected from surrounding objects on the shadow, must be in proportion to the intensity of the
light, the stronger the light, the more the shadow
will be illumined.

MRS. B.

Your remark is perfectly just; but as we have
no means of estimating the degrees of light and of
darkness but by comparison, the strongest light
will appear to produce the deepest shadow. Hence
a total eclipse of the sun occasions a more sensible darkness than mid-night, as it is immediately
contrasted with the strong light of noon-day.

P 3

CAROLINE.

The re-appearance of the sun after an eclipse, must by the same contrast be remarkably brilliant.

MRS. B.

Certainly. There are several things to be observed in regard to the form and extent of shadows. If the luminous body A (fig. 3.) is larger than the opaque body B, the shadow will gradually diminish in size, till it terminate in a point.

CAROLINE.

This is the case with the shadows of the earth and the moon, as the sun which illumines them, is larger than either of those bodies. And why is it not the case with the shadows of terrestrial objects, which are equally illumined by the sun? but their shadows, far from diminishing, are always larger than the object, and increase with the distance from it.

MRS. B.

In estimating the effect of shadows, we must consider the *apparent* not the *real* dimensions of the luminous body; and in this point of view, the sun is a small object compared with the generality of the terrestrial bodies which it illumines: and when the luminous body is less than the opaque body, the shadow will increase with the distance to infinity. All objects, therefore, which are apparently larger than the sun, cast a magnified shadow. This will be best exemplified, by observing the shadow of an object lighted by a candle.

#### EMILY.

I have often noticed, that the shadow of my figure against the wall, grows larger as it is more distant from me, which is owing, no doubt, to the candle that shines on me being much smaller than myself?

#### MRS. B.

Yes. The shadow of a figure A, (fig. 4.) varies in size, according to the distance of the several surfaces B C D E, on which it is described.

#### CAROLINE.

I have observed, that two candles produce two shadows from the same object; whilst it would appear, from what you said, that they should rather produce only half a shadow, that is to say, a very faint one.

#### MRS. B.

The number of lights (in different directions)

while it decreases the intensity of the shadow, increases their number, which always corresponds
with that of the lights; for each light makes the
opaque body cast a different shadow, as illustrated
by figure 5. It represents a ball A, lighted by
three candles B, C, D, and you observe the light
B produces the shadow *b*, the light C the shadow
*c*, and the light D the shadow *d*.

EMILY.

I think we now understand the nature of shadows
very well; but pray what becomes of the rays of
light which opaque bodies arrest in their course,
and the interruption of which is the occasion of
shadows?

MRS. B.

Your question leads to a very important property
of light, *Reflection.* When rays of light encounter
an opaque body, which they cannot traverse, part
of them are absorbed by it, and part are reflected,
and rebound just as an elastic ball which is struck
against a wall.

EMILY.

And is light in its reflection governed by the
same laws as solid elastic bodies?

MRS. B.

Exactly. If a ray of light fall perpendicularly

on an opaque body, it is reflected back in the same line, towards the point whence it proceeded. If it fall obliquely, it is reflected obliquely, but in the opposite direction; the angle of incidence being equal to the angle of reflection. You recollect that law in mechanics?

EMILY.

Oh yes, perfectly.

MRS. B.

If you will shut the shutters, we shall admit a ray of the sun's light through a very small aperture, and I can show you how it is reflected. I now hold this mirror, so that the ray shall fall perpendicularly upon it.

CAROLINE.

I see the ray which falls upon the mirror, but not that which is reflected by it.

MRS. B.

Because its reflection is directly retrograde. The ray of incidence and that of reflection both being in the same line, though in opposite directions, are confounded together.

EMILY.

The ray then which appears to us single, is really

P 5

double, and is composed of the incident ray proceeding to the mirror, and of the reflected ray returning from the mirror.

MRS. B.

Exactly so. We shall now separate them, by
holding the mirror M, (fig. 6.) in such a manner,
that the incident ray A B shall fall obliquely upon
it — you see the reflected ray B C, is marching
off in another direction. If we draw a line from the
point of incidence B, perpendicular to the mirror,
it will divide the angle of incidence from the angle
of reflection, and you will see that they are equal.

EMILY.

Exactly; and now that you hold the mirror so,
that the ray falls more obliquely on it, it is also reflected more obliquely, preserving the equality of
the angles of incidence and reflection.

MRS. B.

It is by reflected rays only that we see opaque
objects. Luminous bodies send rays of light immediately to our eyes, but the rays which they send
to other bodies are invisible to us, and are seen
only when they are reflected or transmitted by those
bodies to our eyes.

EMILY.

But have we not just seen the ray of light in its

passage from the sun to the mirror, and its reflection? yet in neither case were those rays in a direction to enter our eyes.

### MRS. B.

No. What you saw was the light reflected to your eyes by small particles of dust floating in the air, and on which the ray shone in its passage to and from the mirror.

### CAROLINE.

Yet I see the sun shining on that house yonder,, as clearly as possible.

### MRS. B.

Indeed you cannot see a single ray which passes from the sun to the house; you see no rays but those which enter your eyes; therefore it is the rays which are reflected by the house to you, and not those which proceed from the sun to the house, that are visible to you.

### CAROLINE.

Why then does one side of the house appear to be in sunshine, and the other in the shade? for if I cannot see the sun shine upon it, the whole of the house should appear in the shade.

## MRS. B.

That side of the house which the sun shines upon, reflects more vivid and luminous rays than the side which is in shadow, for the latter is illumined only by rays reflected upon it by other objects, these rays are therefore twice reflected before they reach your sight; and as light is more or less absorbed by the bodies it strikes upon, every time a ray is reflected its intensity is diminished.

## CAROLINE.

Still I cannot reconcile myself to the idea, that we do not see the sun's rays shining on objects, but only those which objects reflect to us.

## MRS. B.

I do not, however, despair of convincing you of it. Look at that large sheet of water, can you tell me why the sun appears to shine on one part of it only?

## CAROLINE.

No, indeed; for the whole of it is equally exposed to the sun. This partial brilliancy of water has often excited my wonder; but it has struck me more particularly by moon-light. I have frequently observed a vivid streak of moonshine on the sea, while the rest of the water remained in deep obscurity, and yet there was no apparent obstacle to prevent the moon from shining on every part of the water equally.

By moon-light the effect is more remarkable, on account of the deep obscurity of the other parts of the water; while by the sun's light the effect is too strong for the eye to be able to contemplate it.

But if the sun really shines on every part of that sheet of water, why does not every part of it reflect rays to my eyes?

The reflected rays are not attracted out of their natural course by your eyes. The direction of a reflected ray, you know, depends on that of the incident ray; the sun's rays, therefore, which fall with various degrees of obliquity upon the water, are reflected in directions equally various; some of these will meet your eyes, and you will see them, but those which fall elsewhere are invisible to you.

The streak of sunshine, then, which we now see upon the water, is composed of those rays which by their reflection happen to fall upon my eyes?

MRS. B.

Precisely.

EMILY.

But is that side of the house yonder, which ap
pears to be in shadow, really illumined by the sun,
and its rays reflected another way?

MRS. B.

No; that is a different case from the sheet of
water. That side of the house is really in shadow;
it is the west side, which the sun cannot shine upon
till the afternoon.

EMILY.

Those objects, then, which are illumined by reflected rays, and those which receive direct rays
from the sun, but which do not reflect those
rays towards us, appear equally in shadow?

MRS. B.

Certainly; for we see them both illumined by
reflected rays. That part of the sheet of water, over
which the trees cast a shadow, by what light do
you see it?

EMILY.

Since it is not by the sun's direct rays, it must
be by those reflected on it from other objects, and
which it again reflects to us.

CAROLINE.

But if we see all terrestrial objects by reflected light, (as we do the moon,) why do they appear so bright and luminous? I should have supposed, that reflected rays would have been dull and faint, like those of the moon.

MRS. B.

The moon reflects the sun's light with as much vividness as any terrestrial object. If you look at it on a clear night, it will appear as bright as a sheet of water, the walls of a house, or any object seen by day-light, and on which the sun shines. The rays of the moon are doubtless feeble, when compared with those of the sun; but that would not be a fair comparison, for the former are incident, the latter reflected rays.

CAROLINE.

True; and when we see terrestrial objects by moon-light, the light has been twice reflected, and is consequently proportionally fainter.

MRS. B.

In traversing the atmosphere, the rays, both of the sun and moon, lose some of their light. For though the pure air is a transparent medium, which transmits the rays of light freely, we have observed, that near the surface of the earth it is loaded with vapours and exhalations, by which some portion of them are absorbed.

I have often noticed, that an object on the summit of a hill appears more distinct than one at an equal distance in a valley, or on a plain; which is owing, I suppose, to the air being more free from vapours in an elevated situation, and the reflected rays being consequently brighter.

That may have some sensible effect; but when an object on the summit of a hill has a back ground of light sky, the contrast with the object makes its outline more distinct.

I now feel well satisfied, that we see opaque objects only by reflected rays; but I do not understand how these rays show us the objects from which they proceed?

The rays of light enter at the pupil of the eye, and proceed to the retina, or optic nerve, which is situated at the back part of the eye-ball; and there they describe the figure, colour, and (excepting size) form a perfect representation of the object

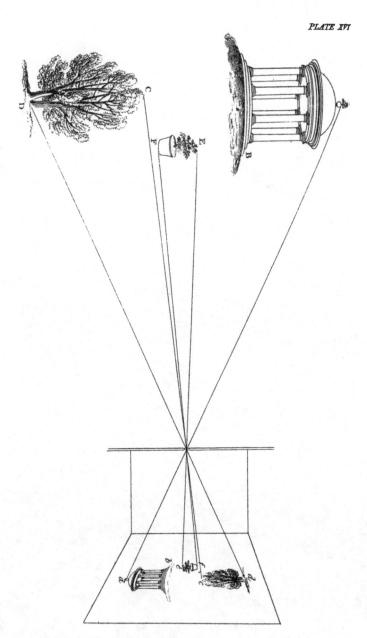

PLATE XVI

Published by Longman & Cᵒ June 1.ˢᵗ 1819.

Lowry Sc.

from which they proceed. We shall again close the shutters, and admit the light through the small aperture, and you will see a picture on the wall, opposite the aperture, similar to that which is delineated on the retina of the eye.

CAROLINE.

Oh, how wonderful! There is an exact picture in miniature of the garden, the gardener at work, the trees blown about by the wind. The landscape would be perfect, if it were not reversed; the ground being above, and the sky beneath.

MRS. B.

It is not enough to admire, you must understand this phenomenon, which is called a camera obscura, from the necessity of darkening the room, in order to exhibit it.

This picture is produced by the rays of light reflected from the various objects in the garden, and which are admitted through the hole in the window-shutter.

The rays from the glittering weathercock at the top of the alcove A, (Plate XVI. fig. 1.) represent it in this spot *a*; for the weathercock being much higher than the aperture in the shutter, only a few of the rays, which are reflected by it in an obliquely descending direction, can find entrance there. The rays of light, you know, always move in

straight lines; those, therefore, which enter the
room in a descending direction, will continue their
course in the same direction, and will, consequently,
fall upon the lower part of the wall opposite the
aperture, and represent the weathercock reversed in
that spot, instead of erect in the uppermost part of
the landscape.

### EMILY.

And the rays of light from the steps (B) of the
alcove, in entering the aperture, ascend, and will
describe those steps in the highest instead of the
lowest part of the landscape.

### MRS. B.

Observe, too, that the rays coming from the
alcove, which is to our left, describe it on the wall
to the right; while those which are reflected by the
walnut-tree C D, to our right, delineate its figure
in the picture to the left, c d. Thus the rays,
coming in different directions, and proceeding
always in right lines, cross each other at their entrance through the aperture: those which come
above proceed below, those from the right go to
the left, those from the left towards the right; thus
every object is represented in the picture, as occupying a situation the very reverse of that which it
does in nature.

CAROLINE.

Excepting the flower-pot E F, which, though its position is reversed, has not changed its situation in the landscape.

MRS. B.

The flower-pot is directly in front of the aperture; so that its rays fall perpendicularly upon it, and, consequently, proceed perpendicularly to the wall, where they delineate the object directly behind the aperture.

EMILY.

And is it thus that the picture of objects is painted on the retina of the eye?

MRS. B.

Precisely.   The pupil of the eye, through which the rays of light enter, represents the aperture in the window-shutter; and the image delineated on the retina, is exactly similar to the picture on the wall.

CAROLINE.

You do not mean to say, that we see only the representation of the object which is painted on the retina, and not the object itself?

MRS. B.

If, by sight, you understand that sense by which

the presence of objects is perceived by the mind,
through the means of the eyes, we certainly see
only the image of those objects painted on the
retina.

CAROLINE.

This appears to me quite incredible.

MRS. B.

The nerves are the only part of our frame capable of sensation: they appear, therefore, to be the
instruments which the mind employs in its perceptions; for a sensation always conveys an idea to
the mind.  Now it is known, that our nerves can
be affected only by contact; and for this reason the
organs of sense cannot act at a distance: for instance, we are capable of smelling only particles
which are actually in contact with the nerves of the
nose.  We have already observed, that the odour
of a flower consists in effluvia, composed of very
minute particles, which penetrate the nostrils, and
strike upon the olfactory nerves, which instantly
convey the idea of smell to the mind.

EMILY.

And sound, though it is said to be heard at a
distance, is, in fact, heard only when the vibrations
of the air, which convey it to our ears, strike upon
the auditory nerve.

CAROLINE.

There is no explanation required, to prove that the senses of feeling and of tasting are excited only by contact.

MRS. B.

And I hope to convince you, that the sense of sight is so likewise. The nerves, which constitute the sense of sight, are not different in their nature from those of the other organs; they are merely instruments which convey ideas to the mind, and can be affected only on contact. Now, since real objects cannot be brought to touch the optic nerve, the image of them is conveyed thither by the rays of light proceeding from real objects, which actually strike upon the optic nerve, and form that image which the mind perceives.

CAROLINE.

While I listen to your reasoning, I feel convinced; but when I look upon the objects around, and think that I do not see them, but merely their image painted in my eyes, my belief is again staggered. I cannot reconcile myself to the idea, that I do not really see this book which I hold in my hand, nor the words which I read in it.

MRS. B.

Did it ever occur to you as extraordinary, that you never beheld your own face?

### CAROLINE.

No; because I so frequently see an exact representation of it in the looking-glass.

### MRS. B.

You see a far more exact representation of objects on the retina of your eye: it is a much more perfect mirror than any made by art.

### EMILY.

But is it possible, that the extensive landscape, which I now behold from the window, should be represented on so small a space as the retina of the eye?

### MRS. B.

It would be impossible for art to paint so small and distinct a miniature; but nature works with a surer hand, and a more delicate pencil. That power, which forms the feathers of the butterfly, and the flowerets of the daisy, can alone pourtray so admirable and perfect a miniature as that which is represented on the retina of the eye.

### CAROLINE.

But, Mrs. B., if we see only the image of objects, why do we not see them reversed, as you showed us they were in the camera obscura? Is not that a strong argument against your theory?

##### MRS. B.

Not an unanswerable one, I hope. The image on the retina, it is true, is reversed, like that in the camera obscura; as the rays, unless from a very small object, intersect each other on entering the pupil, in the same manner as they do on entering the camera obscura. The scene, however, does not excite the idea of being inverted, because we always see an object in the direction of the rays which it sends to us.

##### EMILY.

I confess I do not understand that.

##### MRS. B.

It is, I think, a difficult point to explain clearly. A ray which comes from the upper part of an object, describes the image on the lower part of the retina; but experience having taught us, that the direction of that ray is from above, we consider that part of the object it represents as uppermost. The rays proceeding from the lower part of an object fall upon the upper part of the retina; but as we know their direction to be from below, we see that part of the object they describe as the lowest.

##### CAROLINE.

When I want to see an object above me, I look up; when an object below me, I look down. Does not this prove that I see the objects themselves? for if I beheld only the image, there would be no necessity for looking up or down, according as the object was higher or lower than myself.

MRS. B.

I beg your pardon. When you look up to an elevated object, it is in order that the rays reflected from it should fall upon the retina of your eyes; but the very circumstance of directing your eyes upwards convinces you that the object is elevated, and teaches you to consider as uppermost the image it forms on the retina, though it is, in fact, represented in the lowest part of it. When you look down upon an object, you draw your conclusion from a similar reasoning; it is thus that we see all objects in the direction of the rays which reach our eyes.

But I have a further proof in favour of what I have advanced, which, I hope, will remove your remaining doubts; I shall, however, defer it till our next meeting, as the lesson has been sufficiently long to-day.

# CONVERSATION XV.

## OPTICS — *continued.*

## ON THE ANGLE OF VISION, AND THE REFLECTION OF MIRRORS.

ANGLE OF VISION. — REFLECTION OF PLAIN MIRRORS. — REFLECTION OF CONVEX MIRRORS. — REFLECTION OF CONCAVE MIRRORS.

### CAROLINE.

WELL, Mrs. B., I am very impatient to hear what further proofs you have to offer in support of your theory. You must allow that it was rather provoking to dismiss us as you did at our last meeting.

### MRS. B.

You press so hard upon me with your objections, that you must give me time to recruit my forces.

Q

Can you tell me, Caroline, why objects at a distance appear smaller than they really are?

I know no other reason than their distance.

I do not think I have more cause to be satisfied with your reasons, than you appear to be with mine. We must refer again to the camera obscura to account for this circumstance; and you will find, that the different apparent dimensions of objects at different distances, proceed from our seeing, not the objects themselves, but merely their image on the retina. Fig. 1. Plate XVII. represents a row of trees, as viewed in the camera obscura. I have expressed the direction of the rays, from the objects to the image, by lines. Now, observe, the ray which comes from the top of the nearest tree, and that which comes from the foot of the same tree, meet at the aperture, forming an angle of about twenty-five degrees; this is called the angle of vision, under which we see the tree. These rays cross each other at the aperture, forming equal angles on each side of it, and represent the tree inverted in the camera obscura. The degrees of the image are considerably smaller than those of the object, but the proportions are perfectly preserved.

Now let us notice the upper and lower ray, from

the most distant tree; they form an angle of not more than twelve or fifteen degrees, and an image of proportional dimensions. Thus, two objects of the same size, as the two trees of the avenue, form figures of different sizes in the camera obscura, according to their distance; or, in other words, according to the angle of vision under which they are seen. Do you understand this?

CAROLINE.

Perfectly.

MRS. B.

Then you have only to suppose that the representation in the camera obscura is similar to that on the retina.

Now since objects of the same magnitudes appear to be of different dimensions, when at different distances from us, let me ask you, which it is that we see; the real objects, which we know do not vary in size, or the images, which we know do vary according to the angle of vision under which we see them?

CAROLINE.

I must confess, that reason is in favour of the latter. But does that chair at the further end of the room form an image on my retina much smaller

Q 2

than this which is close to me? they appear exactly of the same size.

I assure you they do not. The experience we acquire by the sense of touch corrects the errors of our sight with regard to objects within our reach. You are so perfectly convinced of the real size of objects which you can handle, that you do not attend to their apparent difference.

Does that house appear to you much smaller than when you are close to it?

No, because it is very near us.

And yet you can see the whole of it through one of the windows of this room. The image of the house on your retina must, therefore, be smaller than that of the window through which you see it. It is your knowledge of the real size of the house which prevents your attending to its apparent magnitude. If you were accustomed to draw from nature, you would be fully aware of this difference.

And pray, what is the reason that, when we look

up an avenue, the trees not only appear smaller as they are more distant, but seem gradually to approach each other till they meet in a point?

MRS. B.

Not only the trees, but the road which separates the two rows, forms a smaller visual angle, in proportion as it is more distant from us; therefore the width of the road gradually diminishes as well as the size of the trees, till at length the road apparently terminates in a point, at which the trees seem to meet.

But this effect of the angle of vision will be more fully illustrated by a little model of an avenue, which I have made for that purpose. It consists of six trees, leading to a hexagonal temple, and viewed by an eye, on the retina of which the picture of the objects is delineated.

I beg that you will not criticise the proportions; for though the eye is represented the size of life, while the trees are not more than three inches high, the disproportion does not affect the principle, which the model is intended to elucidate.

EMILY.

The threads which pass from the objects through the pupil of the eye to the retina, are, I suppose, to represent the rays of light which convey the image of the objects to the retina?

Q 3

**MRS. B.**

Yes. I have been obliged to limit the rays to a very small number, in order to avoid confusion; there are, you see, only two from each tree.

**CAROLINE.**

But as one is from the summit, and the other from the foot of the tree, they exemplify the different angles under which we see objects at different distances, better than if there were more.

**MRS. B.**

There are seven rays proceeding from the temple, one from the summit, and two from each of the angles that are visible to the eye, as it is situated; from these you may form a just idea of the difference of the angle of vision of objects viewed obliquely, or in front; for though the six sides of the temple are of equal dimensions, that which is opposite to the eye is seen under a much larger angle, than those which are viewed obliquely. It is on this principle that the laws of perspective are founded.

**EMILY.**

I am very glad to know that, for I have lately begun to learn perspective, which appeared to me a very dry study; but now that I am acquainted with the principles on which it is founded, I shall find it much more interesting.

In drawing a view from nature, then, we do not copy the real objects, but the image they form on the retina of our eyes?

MRS. B.

Certainly. In sculpture, we copy nature as she really exists; in painting, we represent her as she appears to us. It was on this account that I found it difficult to explain by a drawing the effects of the angle of vision, and was under the necessity of constructing a model for that purpose.

EMILY.

I hope you will allow us to keep this model some time, in order to study it more completely, for a great deal may be learned from it; it illustrates the nature of the angle of vision, the apparent diminution of distant objects, and the inversion of the image on the retina. But pray, why are the threads that represent the rays of light, coloured, the same as the objects from which they proceed?

MRS. B.

That is a question which you must excuse my

Q 4

answering at present, but I promise to explain it to you in due time.

I consent very willingly to your keeping the model, on condition that you will make an imitation of it, on the same principle, but representing different objects.

We must now conclude the observations that remain to be made on the angle of vision.

If an object, with an ordinary degree of illumination, does not subtend an angle of more than two seconds of a degree, it is invisible. There are consequently two cases in which objects may be invisible, either if they are too small, or so distant as to form an angle less than two seconds of a degree.

In like manner, if the velocity of a body does not exceed 20 degrees in an hour, its motion is imperceptible.

CAROLINE.

A very rapid motion may then be imperceptible, provided the distance of the moving body is sufficiently great.

MRS. B.

Undoubtedly; for the greater its distance, the smaller will be the angle under which its motion will appear to the eye. It is for this reason that the motion of the celestial bodies is invisible, notwithstanding their immense velocity.

EMILY.

I am surprised that so great a velocity as 20 degrees an hour should be invisible.

MRS. B.

The real velocity depends altogether on the space comprehended in each degree; and this space depends on the distance of the object, and the obliquity of its path. Observe, likewise, that we cannot judge of the velocity of a body in motion unless we know its distance; for supposing two men to set off at the same moment from A and B, (fig. 2.) to walk each to the end of their respective lines C and D; if they perform their walk in the same space of time, they must have proceeded at a very different rate, and yet to an eye situated at E, they will appear to have moved with equal velocity: because they will both have gone through an equal number of degrees, though over a very unequal length of ground. Sight is an extremely useful sense no doubt, but it cannot always be relied on, it deceives us both in regard to the size and the distance of objects; indeed our senses would be very liable to lead us into error, if experience did not set us right.

Q 5

### EMILY.

Between the two, I think that we contrive to acquire a tolerably accurate idea of objects.

### MRS. B.

At least sufficiently so for the general purposes of life. · To convince you how requisite experience is to correct the errors of sight, I shall relate to you the case of a young man who was blind from his infancy, and who recovered his sight at the age of fourteen, by the operation of couching. At first, he had no idea either of the size or distance of objects, but imagined that every thing he saw touched his eyes; and it was not till after having repeatedly felt them, and walked from one object to another, that he acquired an idea of their respective dimensions, their relative situations, and their distances.

### CAROLINE.

The idea that objects touched his eyes, is however not so absurd, as it at first appears; for if we consider that we see only the image of objects, this image actually touches our eyes.

### MRS. B.

That is doubtless the reason of the opinion he formed, before the sense of touch had corrected his judgment.

CAROLINE.

But since an image must be formed on the retina of each of our eyes, why do we not see objects double?

MRS. B.

The action of the rays on the optic nerve of each eye is so perfectly similar, that they produce but a single sensation, the mind therefore receives the same idea, from the retina of both eyes, and conceives the object to be single.

CAROLINE.

This is difficult to comprehend, and, I should think, can be but conjectural.

MRS. B.

I can easily convince you, that you have a distinct image of an object formed on the retina of each eye. Look at the bell-rope, and tell me do you see it to the right or the left of the pole of the fire-skreen?

CAROLINE.

A little to the right of it.

MRS. B.

Then shut your right eye, and you will see it to the left of the pole.

That is true indeed !

There are evidently two representations of the bell-rope in different situations, which must be owing to an image of it being formed on both eyes; if the action of the rays therefore on each retina were not so perfectly similar as to produce but one sensation, we should see double, and we find that to be the case with many persons who are afflicted with a disease in one eye, which prevents the rays of light from affecting it, in the same manner as the other.

Pray, Mrs. B., when we see the image of an object in a looking-glass, why is it not inverted as in the camera obscura, and on the retina of the eye?

Because the rays do not enter the mirror by a small aperture, and cross each other, as they do at the orifice of a camera obscura, or the pupil of the eye.

When you view yourself in a mirror, the rays from your eyes fall perpendicularly upon it, and are reflected in the same line; the image is therefore

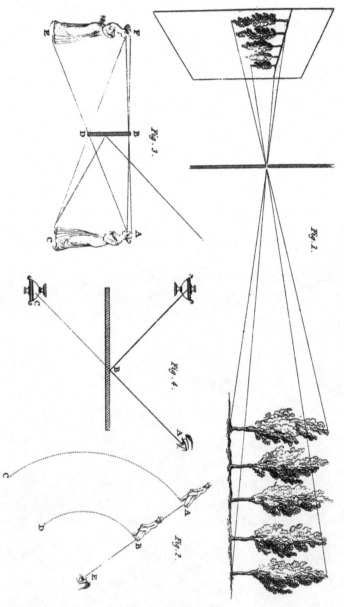

PLATE XVII

*Fig. 3.*

*Fig. 1.*

*Fig. 4.*

*Fig. 2.*

Published by Longman & Co. June 1.ˢᵗ 1819.

described behind the glass, and is situated in the same manner as the object before it.

EMILY.

Yes, I see that it is; but the looking-glass is not nearly so tall as I am, how is it therefore that I can see the whole of my figure in it?

MRS. B.

It is not necessary that the mirror should be more than half your height, in order that you may see the whole of your person in it, (fig. 3.)  The ray of light C D from your eye, which falls perpendicularly on the mirror B D, will be reflected back in the same line; but the ray from your feet will fall obliquely on the mirror, for it must ascend in order to reach it; it will therefore be reflected in the line D A: and since we view objects in the direction of the reflected rays, which reach the eye, and that the image appears at the same distance behind the mirror that the object is before it, we must continue the line A D to E, and the line C D to F, at the termination of which, the image will be represented.

EMILY.

Then I do not understand why I should not see the whole of my person in a much smaller mirror, for a ray of light from my feet would always reach it, though more obliquely.

MRS. B.

True; but the more obliquely the ray falls on the mirror, the more obliquely it will be reflected; the ray would therefore be reflected above your head, and you could not see it. This is shown by the dotted line (fig. 3.)

Now stand a little to the right of the mirror, so that the rays of light from your figure may fall obliquely on it ——

EMILY.

There is no image formed of me in the glass now.

MRS. B.

I beg your pardon, there is; but you cannot see it, because the incident rays falling obliquely on the mirror will be reflected obliquely in the opposite direction, the angles of incidence and of reflection being equal. Caroline, place yourself in the direction of the reflected rays, and tell me whether you do not see Emily's image in the glass?

CAROLINE

Let me consider. — In order to look in the direc-

I I

tion of the reflected rays, I must place myself as much to the left of the glass as Emily stands to the right of it.—Now I see her image, but it is not straight before me, but before her; and appears at the same distance behind the glass, as she is in front of it.

MRS. B.

You must recollect, that we always see objects in the direction of the last rays which reach our eyes. Figure 4. represents an eye looking at the image of a vase, reflected by a mirror; it must see it in the direction of the ray A B, as that is the ray which brings the image to the eye; prolong the ray to C, and in that spot will the image appear.

CAROLINE.

I do not understand why a looking-glass reflects the rays of light; for glass is a transparent body which should transmit them?

MRS. B.

It is not the glass that reflects the rays which form the image you behold, but the mercury behind it. The glass acts chiefly as a transparent case, through which the rays find an easy passage.

CAROLINE.

Why then should not mirrors be made simply of mercury?

MRS. B.

Because mercury is a fluid. By amalgamating it with tin-foil, it becomes of the consistence of paste, attaches itself to the glass, and forms in fact a mercurial mirror, which would be much more perfect without its glass cover, for the purest glass is never perfectly transparent; some of the rays therefore are lost during their passage through it, by being either absorbed, or irregularly reflected.

This imperfection of glass mirrors has introduced the use of metallic mirrors, for optical purposes.

EMILY.

But since all opaque bodies reflect the rays of light, I do not understand why they are not all mirrors?

CAROLINE.

A curious idea indeed, sister; it would be very gratifying to see oneself in every object at which one looked.

MRS. B.

It is very true that all opaque objects reflect

light; but the surface of bodies in general is so rough and uneven, that their reflection is extremely irregular, which prevents the rays from forming an image on the retina. This you will be able to understand better, when I shall explain to you the nature of vision, and the structure of the eye.

You may easily conceive the variety of directions in which rays would be reflected by a nutmeg-grater, on account of the inequality of its surface, and the number of holes with which it is pierced. All solid bodies resemble the nutmeg-grater in these respects, more or less; and it is only those which are susceptible of receiving a polish, that can be made to reflect the rays with regularity. As hard bodies are of the closest texture, the least porous, and capable of taking the highest polish, they make the best mirrors; none therefore are so well calculated for this purpose as metals.

CAROLINE.

But the property of regular reflection is not confined to this class of bodies; for I have often seen myself in a highly polished mahogany table.

MRS. B.

Certainly; but as that substance is less durable, and its reflection less perfect, than that of metals, I believe it would seldom be chosen for the purpose of a mirror.

There are three kinds of mirrors used in optics; the plain or flat, which are the common mirrors we have just mentioned; convex mirrors; and concave mirrors. The reflection of the two latter is very different from that of the former. The plain mirror, we have seen, does not alter the direction of the reflected rays, and forms an image behind the glass exactly similar to the object before it. A convex mirror has the peculiar property of making the reflected rays diverge, by which means it diminishes the image; and a concave mirror makes the rays converge, and under certain circumstances, magnifies the image.

We have a convex mirror in the drawing-room, which forms a beautiful miniature picture of the objects in the room; and I have often amused myself with looking at my magnified face in a concave mirror. But I hope you will explain to us why the one enlarges, while the other diminishes the objects it reflects.

Let us begin by examining the reflection of a convex mirror. This is formed of a portion of the exterior surface of a sphere. When several parallel rays fall upon it, that ray only which, if prolonged, would pass through the centre or axis of the mirror,

PLATE XVIII

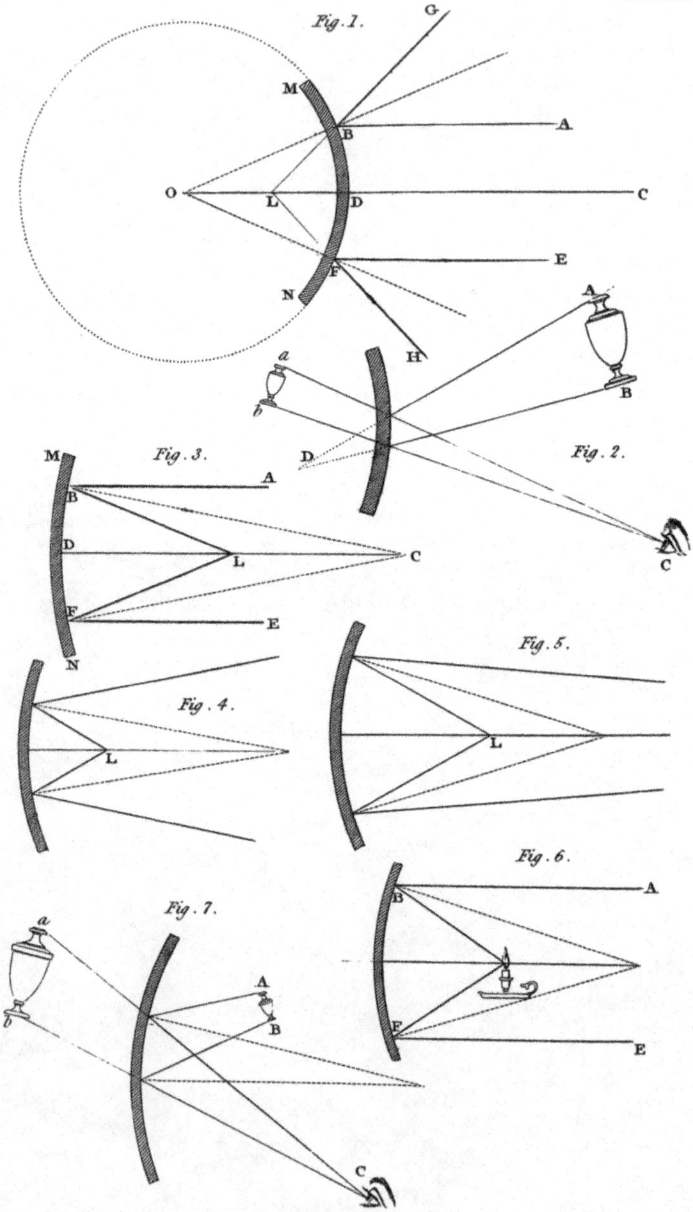

Fig. 1.

Fig. 2.

Fig. 3.

Fig. 4.

Fig. 5.

Fig. 6.

Fig. 7.

Published by Longman & Co. June 1.st 1819.

Lowry Sc.

is perpendicular to it.   In order to avoid confusion,
I have, in fig. 1. Plate XVIII., drawn only three
parallel lines, A B, C D, E F, to represent rays
falling on the convex mirror M N; the middle ray,
you will observe, is perpendicular to the mirror, the
others fall on it obliquely.

### CAROLINE.

As the three rays are parallel, why are they not
all perpendicular to the mirror?

### MRS. B.

They would be so to a flat mirror; but as this is
spherical, no ray can fall perpendicularly upon it
which is not directed towards the centre of the
sphere.

### EMILY.

Just as a weight falls perpendicularly to the earth
w hen gravity attracts it towards the centre.

### MRS. B.

In order, therefore, that rays may fall perpendicularly to the mirror at B and F, the rays must be
in the direction of the dotted lines, which, you may
observe, meet at the centre O of the sphere, of
which the mirror forms a portion.

Now can you tell me in what direction the three rays, A B, C D, E F, will be reflected?

Yes, I think so: the middle ray falling perpendicularly on the mirror, will be reflected in the same line: the two others falling obliquely, will be reflected obliquely to G H; for the dotted lines you have drawn are perpendiculars, which divide their angles of incidence and reflection.

Extremely well, Emily: and since we see objects in the direction of the reflected ray, we shall see the image at L, which is the point at which the reflected rays, if continued through the mirror, would unite and form an image. This point is equally distant from the surface and centre of the sphere, and is called the imaginary focus of the mirror.

Pray, what is the meaning of focus?

A point at which converging rays unite. And it is in this case called an imaginary focus; because the rays do not really unite at that point, but only appear to do so: for the rays do not pass through the mirror, since they are reflected by it.

I do not yet understand why an object appears smaller when viewed in a convex mirror.

It is owing to the divergence of the reflected rays. You have seen that a convex mirror converts, by reflection, parallel rays into divergent rays; rays that fall upon the mirror divergent, are rendered still more so by reflection, and convergent rays are reflected either parallel, or less convergent. If then an object be placed before any part of a convex mirror, as the vase A B, fig. 2. for instance, the two rays from its extremities, falling convergent on the mirror, will be reflected less convergent, and will not come to a focus till they arrive at C; then an eye placed in the direction of the reflected rays, will see the image formed in (or rather behind) the mirror at *a b*,

But the reflected rays do not appear to me to converge less than the incident rays. I should have supposed that, on the contrary, they converged more, since they meet in a point?

They would unite sooner than they actually do, if they were not less convergent than the incident rays: for observe, that if the incident rays, instead

of being reflected by the mirror, continued their
course in their original direction, they would come
to a focus at D, which is considerably nearer to the
mirror than at C; the image is therefore seen under
a smaller angle than the object; and the more distant the latter is from the mirror, the less is the
image reflected by it.

You will now easily understand the nature of the
reflection of concave mirrors.   These are formed of
a portion of the internal surface of a hollow sphere,
and their peculiar property is to converge the rays
of light.

Can you discover, Caroline, in what direction
the three parallel rays, A B, C D, E F, which fall
on the concave mirror M N, (fig. 3.) are reflected?

CAROLINE.

I believe I can.   The middle ray is sent back
in the same line, as it is in the direction of the axis
of the mirror; and the two others will be reflected
obliquely, as they fall obliquely on the mirror.   I
must now draw two dotted lines perpendicular to
their points of incidence, which will divide their
angles of incidence and reflection; and in order that
those angles may be equal, the two oblique rays
must be reflected to L, where they will unite with
the middle ray.

MRS. B.

Very well explained.   Thus you see, that when

any number of parallel rays fall on a concave mirror, they are all reflected to a focus: for in proportion as the rays are more distant from the axis of the mirror, they fall more obliquely upon it, and are more obliquely reflected; in consequence of which they come to a focus in the direction of the axis of the mirror, at a point equally distant from the centre and the surface of the sphere, and this point is not an imaginary focus, as happens with the convex mirror, but is the true focus at which the rays unite.

EMILY.

Can a mirror form more than one focus by reflecting rays?

MRS. B.

Yes.   If rays fall convergent on a concave mirror, (fig. 4.) they are sooner brought to a focus, L, than parallel rays; their focus is therefore nearer to the mirror M N.   Divergent rays are brought to a more distant focus than parallel rays, as in figure 5, where the focus is at L; but the true focus of mirrors, either convex or concave, is that of parallel rays, which is equally distant from the centre, and the surface of the sphere.

I shall now show you the reflection of real rays of light, by a metallic concave mirror.   This is one made of polished tin, which I expose to the sun, and as it shines bright, we shall be able to collect the rays into a very brilliant focus. I hold a piece of paper where I imagine the focus to be situated; you may see by the vivid spot of light on the paper, how much the rays converge: but it is not yet exactly in the focus; as I approach the paper to that point, observe how the brightness of the spot of light increases, while its size diminishes.

### CAROLINE.

That must be occasioned by the rays becoming closer together. I think you hold the paper just in the focus now, the light is so small and dazzling — Oh, Mrs. B., the paper has taken fire !

### MRS. B.

The rays of light cannot be concentrated, without, at the same time, accumulating a proportional quantity of heat: hence concave mirrors have obtained the name of burning-mirrors.

### EMILY.

I have often heard of the surprising effects of burning-mirrors, and I am quite delighted to understand their nature.

### CAROLINE.

It cannot the be true focus of the mirror at

which the rays of the sun unite, for as they proceed from a point, they must fall divergent upon the mirror.

MRS. B.

Strictly speaking, they certainly do.   But when rays come from such an immense distance as the sun, their divergence is so trifling, as to be imperceptible; and they may be considered as parallel: their point of union is, therefore, the true focus of the mirror, and there the image of the object is represented.

Now that I have removed the mirror out of the influence of the sun's rays, if I place a burning taper in the focus, how will its light be reflected? (Fig. 6.)

CAROLINE.

That, I confess, I cannot say.

MRS. B.

The ray which falls in the direction of the axis of the mirror, is reflected back in the same line; but let us draw two other rays from the focus, falling on the mirror at B and F; the dotted lines are perpendicular to those points, and the two rays will therefore be reflected to A and E.

R

CAROLINE.

Oh, now I understand it clearly. The rays which proceed from a light placed in the focus of a concave mirror fall divergent upon it, and are reflected parallel. It is exactly the reverse of the former experiment, in which the sun's rays fell parallel on the mirror, and were reflected to a focus.

MRS. B.

Yes: when the incident rays are parallel, the reflected rays converge to a focus; when, on the contrary, the incident rays proceed from the focus, they are reflected parallel. This is an important law of optics, and since you are now acquainted with the principles on which it is founded, I hope that you will not forget it.

CAROLINE.

I am sure that we shall not. But, Mrs. B., you said that the image was formed in the focus of a concave mirror; yet I have frequently seen glass concave mirrors, where the object has been represented within the mirror, in the same manner as in a convex mirror.

MRS. B.

That is the case only, when the object is placed between the mirror and its focus; the image then appears magnified behind, or, as you call it, within the mirror.

CONVER____ XVI.

CAROLINE.

I do not understand why the image should be larger than the object.

MRS. B.

It proceeds from the convergent property of the concave mirror. If an object, A B, (fig. 7.) be placed between the mirror and its focus, the rays from its extremities fall divergent on the mirror, and on being reflected, become less divergent, as if they proceeded from C: to an eye placed in that situation the image will appear magnified behind the mirror at a b, since it is seen under a larger angle than the object.

You now, I hope, understand the reflection of light by opaque bodies. At our next meeting, we shall enter upon another property of light, no less interesting, which is called *refraction*.

# CONVERSATION XVI.

## ON REFRACTION AND COLOURS.

TRANSMISSION OF LIGHT BY TRANSPARENT BO-
DIES. — REFRACTION. — REFRACTION OF THE
ATMOSPHERE. — REFRACTION OF A LENS. — RE-
FRACTION OF THE PRISM. — OF THE COLOURS OF
RAYS OF LIGHT. — OF THE COLOURS OF BODIES.

MRS. B.

THE refraction of light will furnish the subject of to day's lesson.

CAROLINE.

That is a property of which I have not the faintest idea.

MRS. B.

It is the effect which transparent mediums produce on light in its passage through them. Opaque

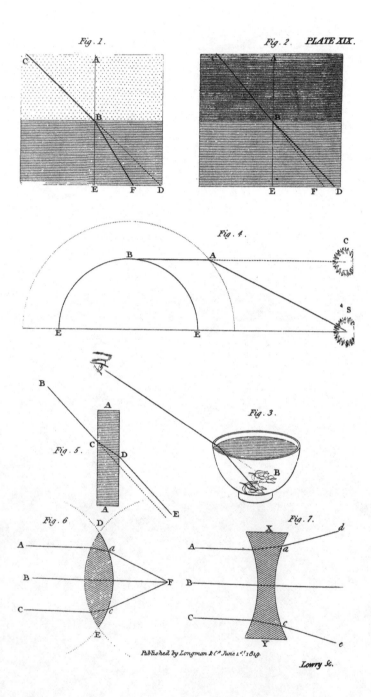

*Fig. 1.*      *Fig. 2.*   **PLATE XIX.**

*Fig. 4.*

*Fig. 3.*

*Fig. 5.*

*Fig. 6.*

*Fig. 7.*

Published by Longman & Co. June 1.ˢᵗ 1819.

bodies, you know, reflect the rays, and transparent bodies transmit them; but it is found, that if a ray, in passing from one medium into another of different density, fall obliquely, it is turned out of its course.

CAROLINE.

It must then be acted on by some new power, otherwise it would not deviate from its first direction.

MRS. B.

The power which causes the deviation of the ray appears to be the attraction of the denser medium. Let us suppose the two mediums to be air and water; if a ray of light passes from air into water, it is more strongly attracted by the latter on account of its superior density.

EMILY.

In what direction does the water attract the ray?

MRS. B.

It must attract it perpendicularly towards it, in the same manner as gravity acts on bodies.

If then a ray A B, (fig. 1. Plate XIX.) fall perpendicularly on water, the attraction of the water acts in the same direction as the course of the ray;

R 3

it will not therefore cause a deviation, and the ray will proceed straight on to E. But if it fall obliquely, as the ray C B, the water will attract it out of its course. Let us suppose the ray to have approached the surface of a denser medium, and that it there begins to be affected by its attraction; this attraction, if not counteracted by some other power, would draw it perpendicularly to the water, at B; but it is also impelled by its projectile force, which the attraction of the denser medium cannot overcome; the ray, therefore, acted on by both these powers, moves in a direction between them, and instead of pursuing its original course to D, or being implicitly guided by the water to E, proceeds towards F, so that the ray appears bent or broken.

CAROLINE.

I understand that very well; and is not this the reason that oars appear bent in water?

MRS. B.

It is owing to the refraction of the rays reflected by the oar; but this is in passing from a dense to a rare medium, for you know that the rays, by means of which you see the oar, pass from water into air.

EMILY.

But I do not understand why a refraction takes

place when a ray passes from a dense into a rare medium; I should suppose that it would be rather less, than more, attracted by the latter.

And it is precisely on that account that the ray is refracted. C B, fig. 2. represents a ray passing obliquely from glass into water: glass being the denser medium, the ray will be more strongly attracted by that which it leaves than by that which it enters. The attraction of the glass acts in the direction A B, while the impulse of projection would carry the ray to F; it moves, therefore, between these directions towards D.

So that a contrary refraction takes place when a ray passes from a dense into a rare medium.

But does not the attraction of the denser medium affect the ray before it touches it?

The distance at which the attraction of the denser medium acts upon a ray is so small as to be insensible; it appears therefore to be refracted only at the point at which it passes from one medium to the other.

Now that you understand the principle of refraction, I will show you the refraction of a real ray of light.   Do you see the flower painted at the bottom of the inside of this tea-cup?   (Fig. 3.)

EMILY.

Yes. — But now you have moved it just out of sight, the rim of the cup hides it.

MRS. B.

Do not stir.   I will fill the cup with water, and you will see the flower again.

EMILY.

I do indeed! Let me try to explain this : when you draw the cup from me so as to conceal the flower, the rays reflected by it no longer met my eyes, but were directed above them; but now that you have filled the cup with water, they are refracted by the attraction of the water, and bent downwards, so as again to enter my eyes.

MRS. B.

You have explained it perfectly: fig. 3. will help to imprint it on your memory.   You must observe that when the flower becomes visible by the refraction of the ray, you do not see it in the situation which it really occupies, but an image of the flower higher in the cup; for as objects always appear to be situated in the direction of the rays which enter the eye, the flower will be seen in the direction of the reflected ray at B.

EMILY.

Then, when we see the bottom of a clear stream of water, the rays which it reflects being refracted in their passage from the water into the air, will make the bottom appear higher than it really is.

MRS. B.

And the water will consequently appear more shallow. Accidents have frequently been occasioned by this circumstance; and boys who are in the habit of bathing should be cautioned not to trust to the apparent shallowness of water, as it will always prove deeper than it appears; unless, indeed, they view it from a boat on the water, which will enable them to look perpendicularly upon it; when the rays from the bottom passing perpendicularly, no refraction will take place.

The refraction of light prevents our seeing the heavenly bodies in their real situation: the light they send to us being refracted in passing into the atmosphere, we see the sun and stars in the direction of the refracted ray; as described in fig. 4. Plate XIX., the dotted line represents the extent of the atmosphere, above a portion of the earth, E B E: a ray of light coming from the sun S falls obliquely on it, at A, and is refracted to B; then, since we see the object in the direction of the refracted ray, a spectator at B will see an image of the sun at C, instead of the real object at S.

EMILY.

But if the sun were immediately over our heads, its rays falling perpendicularly on the atmosphere would not be refracted, and we should then see the real sun, in its true situation.

MRS. B.

You must recollect that the sun is vertical only to the inhabitants of the torrid zone; its rays, therefore, are always refracted in these climates. There is also another obstacle to our seeing the heavenly bodies in their real situations: light, though it moves with extreme velocity, is about eight minutes and an half in its passage from the sun to the earth; therefore, when the rays reach us, the sun must have quitted the spot he occupied on their departure; yet we see him in the direction of those rays, and consequently in a situation which he had abandoned eight minutes and an half before.

EMILY.

When you speak of the sun's motion, you mean,

7

I suppose, his apparent motion, produced by the diurnal motion of the earth?

### MRS. B.

No doubt; the effect being the same, whether it is our earth, or the heavenly bodies which move: it is more easy to represent things as they appear to be, than as they really are.

### CAROLINE.

During the morning, then, when the sun is rising towards the meridian, we must (from the length of time the light is in reaching us) see an image of the sun below that spot which it really occupies.

### EMILY.

But the refraction of the atmosphere counteracting this effect, we may perhaps, between the two, see the sun in its real situation.

### CAROLINE.

And in the afternoon, when the sun is sinking in the west, refraction and the length of time which the light is in reaching the earth, will conspire to render the image of the sun higher than it really is.

### MRS. B.

The refraction of the sun's rays by the atmosphere prolongs our days, as it occasions our seeing

R 6

an image of the sun, both before he rises and after
he sets ; for below the horizon, he still shines
upon the atmosphere, and his rays are thence refracted to the earth.  So likewise we see an image
of the sun before he rises, the rays that previously
fall upon the atmosphere being reflected to the
earth.

### CAROLINE.

On the other hand, we must recollect that light
is eight minutes and an half on its journey; so that,
by the time it reaches the earth, the sun may perhaps be risen above the horizon.

### EMILY.

'Pray do not glass-windows refract the light?

### MRS. B.

They do; but this refraction is not perceptible,
because, in passing through a pane of glass the rays
suffer two refractions, which being in contrary
directions, produce the same effect as if no refraction
had taken place.

### EMILY.

I do not understand that.

### MRS. B.

Fig. 5. Plate XIX. will make it clear to you:

A A represents a thick pane of glass seen edgeways. When the ray B approaches the glass at C, it is refracted by it; and instead of continuing its course in the same direction, as the dotted line describes, it passes through the pane to D; at that point returning into the air, it is again refracted by the glass, but in a contrary direction to the first refraction, and in consequence proceeds to E. Now you must observe that the ray B C and the ray D E being parallel, the light does not appear to have suffered any refraction.

EMILY.

So that the effect which takes place on the ray entering the glass, is undone on its quitting it. Or, to express myself more scientifically, when a ray of light passes from one medium into another, and through that into the first again, the two refractions being equal and in opposite directions, no sensible effect is produced.

MRS. B.

This is the case when the two surfaces of the refracting medium are parallel to each other; if they are not, the two refractions may be made in the same direction, as I shall show you.

When parallel rays (fig. 6.) fall on a piece of glass having a double convex surface, and which is called a *Lens*, that only which falls in the direction of the axis of the lens is perpendicular to the surface; the other rays falling obliquely are refracted towards the axis, and will meet at a point beyond the lens, called its focus.

Of the three rays, A B C, which fall on the lens D E, the rays A and C are refracted in their passage through it, to *a*, and *c*, and on quitting the lens they undergo a second refraction in the same direction which unites them with the ray B, at the focus F.

And what is the distance of the focus from the surface of the lens?

The focal distance depends both upon the form of the lens, and of the refractive power of the substance of which it is made: in a glass lens, both sides of which are equally convex, the focus is situated nearly at the centre of the sphere of which the surface of the lens forms a portion; it is at the distance, therefore, of the radius of the sphere.

There are lenses of various forms, as you will find described in fig. 1. Plate XX. The property of those which have a convex surface is to collect the rays of light to a focus; and of those which have a concave surface, on the contrary, to disperse them. For the rays A C falling on the concave lens X Y,

PLATE XX.

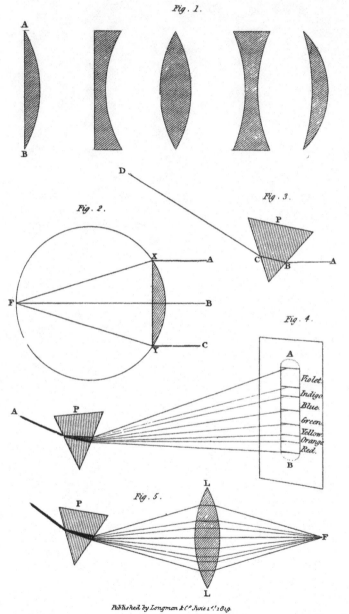

Fig. 1.

Fig. 2.

Fig. 3.

Fig. 4.

Fig. 5.

Published by Longman & Cᵒ. June 1ˢᵗ 1819.

Lowry Sc.

(fig. 7. Plate XIX.) instead of converging towards the ray B, which falls on the axis of the lens, will each be attracted towards the thick edges of the lens, both on entering and quitting it, and will, therefore, by the first refraction, be made to diverge to *a*, *c*, and by the second to *d*, *e*.

CAROLINE.

And lenses which have one side flat and the other convex or concave, as A and B, fig. 1. Plate XX., are, I suppose, less powerful in their refractions?

MRS. B.

Yes; they are called plano-convex, and plano-concave lenses: the focus of the former is at the distance of the diameter of a sphere, of which the convex surface of the lens forms a portion; as represented in fig. 2. Plate XX. The three parallel rays, A B C, are brought to a focus by the plano-convex lens, X Y at F.

I must now explain to you the refraction of a triangular piece of glass, called a prism. (Fig. 3.)

EMILY.

The three sides of this glass are flat; it cannot therefore bring the rays to a focus; nor do I suppose that its refraction will be similar to that of a flat pane of glass, because it has not two sides.

parallel; I cannot therefore conjecture what effect the refraction of a prism can produce.

MRS. B.

The refractions of the light, on entering and on quitting the prism, are both in the same direction. (Fig. 3.)   On entering the prism P, the ray A is refracted from B to C, and on quitting it from C to D.

I will show you this in nature; but for this purpose it will be advisable to close the window-shutters, and admit, through the small aperture, a ray of light, which I shall refract by means of this prism.

CAROLINE.

Oh, what beautiful colours are represented on the opposite wall!   There are all the colours of the rainbow, and with a brightness I never saw equalled. (Fig. 4. Plate XX.)

EMILY.

I have seen an effect, in some respects similar to this, produced by the rays of the sun shining upon glass lustres; but how is it possible that a piece of white glass can produce such a variety of brilliant colours?

MRS. B.

The colours are not formed by the prism,

but existed in the ray previous to its refraction.

CAROLINE.

Yet, before its refraction, it appeared perfectly white.

MRS. B.

The white rays of the sun are composed of coloured rays, which, when blended together, appear colourless or white.

Sir Isaac Newton, to whom we are indebted for the most important discoveries respecting light and colours, was the first who divided a white ray of light, and found it to consist of an assemblage of coloured rays, which formed an image upon the wall, such as you now see exhibited, (fig. 4.) in which are displayed the following series of colours: red, orange, yellow, green, blue, indigo, and violet.

EMILY.

But how does a prism separate these coloured rays?

MRS. B.

By refraction. It appears that the coloured rays have different degrees of refrangibility; in passing through the prism, therefore, they take different directions according to their susceptibility of refraction. The violet rays deviate most from their original course; they appear at one of the ends of the spectrum A B: contiguous to the violet, are the blue rays, being those which have somewhat less refrangibility; then follow, in succession, the green, yellow, orange, and, lastly, the red, which are the least refrangible of the coloured rays.

CAROLINE.

I cannot conceive how these colours, mixed together, can become white?

MRS. B.

That I cannot pretend to explain; but it is a fact that the union of these colours, in the proportions in which they appear in the spectrum, produce in us the idea of whiteness. If you paint a card in compartments with these seven colours, and whirl it rapidly on a pin, it will appear white.

But a more decisive proof of the composition of a white ray is afforded by reuniting these coloured rays, and forming with them a ray of white light.

CAROLINE.

If you can take a ray of white light to pieces, and put it together again, I shall be quite satisfied.

MRS. B.

This can be done by letting the coloured rays, which have been separated by a prism, fall upon a lens, which will converge them to a focus; and if, when thus reunited, we find that they appear white as they did before refraction, I hope that you will be convinced that the white rays are a compound of the several coloured rays.   The prism P, you see, (fig. 5.) separates a ray of white light into seven coloured rays, and the lens L L brings them to a focus at F, where they again appear white.

CAROLINE.

You succeed to perfection: this is indeed a most interesting and conclusive experiment.

EMILY.

Yet, Mrs. B., I cannot help thinking, that there may perhaps be but three distinct colours in the spectrum, red, yellow, and blue; and that the four others may consist of two of these colours blended together; for, in painting, we find that by mixing red and yellow, we produce orange; with different proportions of red and blue, we make violet or any shade of purple; and yellow and blue form green.   Now it is very natural to suppose, that the refraction of a prism may not be so perfect as to separate the coloured rays of light completely, and that those which are contiguous in order of re-frangibility may encroach on each other, and by mixing produce the intermediate colours, orange, green, violet, and indigo.

Your observation is, I believe, neither quite
wrong, nor quite right. Dr. Wollaston, who has refracted light in a more accurate manner than had
been previously done, by receiving a very narrow
line of light on a prism, found that it formed a
spectrum, consisting of rays of four colours only;
but they were not exactly those you have named as
primitive colours, for they consisted of red, green,
blue, and violet. A very narrow line of yellow was
visible, at the limit of the red and green, which
Dr. Wollaston attributed to the overlapping of the
edges of the red and green light.

But red and green mixed together, do not produce yellow?

Not in painting; but it may be so in the primitive rays of the spectrum. Dr. Wollaston observed
that, by increasing the breadth of the aperture by
which the line of light was admitted, the space occupied by each coloured ray in the spectrum was
augmented, in proportion as each portion en-
croached on the neighbouring colour and mixed
with it; so that the intervention of orange and
yellow, between the red and green, is owing, he
supposes, to the mixture of these two colours, and

the blue is blended on the one side with the green, and on the other with the violet, forming the spectrum as it was originally observed by Sir Isaac Newton, and which I have just shown you.

The rainbow, which exhibits a series of colours so analogous to those of the spectrum, is formed by the refraction of the sun's rays in their passage through a shower of rain, every drop of which acts as a prism, in separating the coloured rays as they pass through it.

#### EMILY.

Pray, Mrs. B., cannot the sun's rays be collected to a focus by a lens in the same manner as they are by a concave mirror?

#### MRS. B.

No doubt the same effect is produced by the refraction of a lens as by the reflection of a concave mirror: in the first, the rays pass through the glass and converge to a focus behind it; in the latter, they are reflected from the mirror, and brought to a focus before it. A lens, when used for the purpose of collecting the sun's rays, is called a burning glass. The sun now shines very bright; if we let the rays fall on this lens you will perceive the focus.

#### EMILY.

Oh yes: the point of union of the rays is very

luminous. I will hold a piece of paper in the focus, and see if it will take fire. The spot of light is extremely brilliant, but the paper does not burn?

### MRS. B.

Try a piece of brown paper; — that you see takes fire almost immediately.

### CAROLINE.

This is surprising; for the light appeared to shine more intensely on the white than on the brown paper.

### MRS. B.

The lens collects an equal number of rays to a focus, whether you hold the white or the brown paper there; but the white paper appears more luminous in the focus, because most of the rays, instead of entering into the paper, are reflected by it; and this is the reason that the paper is not burnt: whilst, on the contrary, the brown paper, which absorbs more light than it reflects, soon becomes heated and takes fire.

### CAROLINE.

This is extremely curious; but why should brown paper absorb more rays than white paper?

MRS. B.

I am far from being able to give a satisfactory
answer to that question.   We can form but mere
conjecture on this point; and suppose that the tendency to absorb, or reflect rays, depends on the
arrangement of the minute particles of the body,
and that this diversity of arrangement renders some
bodies susceptible of reflecting one coloured ray,
and absorbing the others; whilst other bodies have
a tendency to reflect all the colours, and others
again, to absorb them all.

EMILY.

And how do you know which colours bodies
have a tendency to reflect; or which to absorb?

MRS. B.

Because a body always appears to ·be of the
colour which it reflects; for, as we see only by reflected rays, it can appear but of the colour of those
rays.

CAROLINE.

But we see all bodies of their own natural colour,
Mrs. B.; the grass and trees, green; the sky, blue;
the flowers, of various hues.

MRS. B.

True; but why is the grass green? — because it

absorbs all except the green rays; it is therefore these only which the grass and trees reflect to our eyes, and which makes them appear green. The sky and flowers, in the same manner, reflect the various colours of which they appear to us; the rose, the red rays; the violet, the blue; the jonquil, the yellow, &c.

### CAROLINE.

But these are the permanent colours of the grass and flowers, whether the sun's rays shine on them or not.

### MRS. B.

Whenever you see those colours, the flowers must be illumined by some light; and light, from whatever source it proceeds, is of the same nature, composed of the various coloured rays, which paint the grass, the flowers, and every coloured object in nature.

### CAROLINE.

But, Mrs. B., the grass is green, and the flowers are coloured, whether in the dark, or exposed to the light?

### MRS. B.

Why should you think so?

#### CAROLINE.
It cannot be otherwise.

#### MRS. B.
A most philosophical reason indeed! But, as I never saw them in the dark, you will allow me to dissent from your opinion.

#### CAROLINE.
What colour do you suppose them to be, then, in the dark?

#### MRS. B.
None at all; or black, which is the same thing. You can never see objects without light. Light is composed of colours, therefore there can be no light without colours; and though every object is black, or without colour in the dark, it becomes coloured, as soon as it becomes visible. It is visible, indeed, but by the coloured rays which it reflects; therefore we can see it only when coloured.

#### CAROLINE.
All you say seems very true, and I know not what to object to it; yet it appears at the same time incredible! What, Mrs. B., are we all as black as negroes, in the dark? you make me shudder at the thought.

s

### MRS. B.

Your vanity need not be alarmed at the idea, as you are certain of never being seen in that state.

### CAROLINE.

That is some consolation, undoubtedly; but what a melancholy reflection it is, that all nature which appears so beautifully diversified with colours should be one uniform mass of blackness!

### MRS. B.

Is nature less pleasing for being coloured, as well as illumined by the rays of light; and are colours less beautiful, for being accidental, rather than essential properties of bodies?

Providence appears to have decorated nature with the enchanting diversity of colours, which we so much admire, for the sole purpose of beautifying the scene, and rendering it a source of pleasurable enjoyment: it is an ornament which embellishes nature, whenever we behold her. What reason is there to regret that she does not wear it when she is invisible?

### EMILY.

I confess, Mrs. B., that I have had my doubts, as well as Caroline, though she has spared me the pains of expressing them; but I have just thought of an experiment, which, if it succeeds, will, I am

sure, satisfy us both.   It is certain, that we cannot
see bodies in the dark, to know whether they have
then any colour.   But we may place a coloured
body in a ray of light, which has been refracted by
a prism; and if your theory is true, the body, of
whatever colour it naturally is, must appear of the
colour of the ray in which it is placed; for since it
receives no other coloured rays, it can reflect no
others.

### CAROLINE.

Oh! that is an excellent thought, Emily; will
you stand the test Mrs. B.?

### MRS. B.

I consent: but we must darken the room, and
admit only the ray which is to be refracted; otherwise, the white rays will be reflected on the body
under trial, from various parts of the room.
With what do you choose to make the experiment?

### CAROLINE.

This rose: look at it, Mrs. B., and tell me whether it is possible to deprive it of its beautiful
colour?

### MRS. B.

We shall see. — I expose it first to the red rays,

s 2

and the flower appears of a more brilliant hue; but observe the green leaves —

They appear neither red nor green; but of a dingy brown with a reddish glow !

They cannot be green, because they have no green rays to reflect; neither are they red, because green bodies absorb most of the red rays. But though bodies, from the arrangement of their particles, have a tendency to absorb some rays, and reflect others, yet it is not natural to suppose, that bodies are so perfectly uniform in their arrangement, as to reflect only pure rays of one colour, and perfectly absorb the others: it is found, on the contrary, that a body reflects, in great abundance, the rays which determine its colour, and the others in a greater or less degree, in proportion as they are nearer or further from its own colour, in the order of refrangibility. The green leaves of the rose, therefore, will reflect a few of the red rays, which, blended with their natural blackness, give them that brown tinge: if they reflected none of the red rays, they would appear perfectly black. Now I shall hold the rose in the blue rays —

CAROLINE.

Oh, Emily, Mrs. B. is right! look at the rose:
it is no longer red, but of a dingy blue colour.

EMILY.

This is the most wonderful of any thing we have
yet learnt. But, Mrs. B., what is the reason that
the green leaves are of a brighter blue than the rose?

MRS. B.

The green leaves reflect both blue and yellow
rays, which produces a green colour. They are now
in a coloured ray, which they have a tendency to
reflect; they, therefore, reflect more of the blue rays
than the rose, (which naturally absorbs that colour,)
and will, of course, appear of a brighter blue.

EMILY.

Yet, in passing the rose through the different colours of the spectrum, the flower takes them more
readily than the leaves.

MRS. B.

Because the flower is of a paler hue. Bodies
which reflect all the rays are white; those which
absorb them all are black: between these extremes,
the body appears lighter or darker, in proportion
to the quantity of rays they reflect or absorb. This

s 3

rose is of a pale red : it approaches nearer to white than black; it therefore reflects rays more abundantly than it absorbs them.

### EMILY.

But if a rose has so strong a tendency to reflect rays, I should have imagined that it would be of a deep red colour.

### MRS. B.

I mean to say, that it has a general tendency to reflect rays. Pale-coloured bodies reflect all the coloured rays to a certain degree, which produces their paleness, approaching to whiteness : but one colour they reflect more than the rest; this predominates over the white, and determines the colour of the body. Since, then, bodies of a pale colour in some degree reflect all the rays of light, in passing through the various colours of the spectrum, they will reflect them all with tolerable brilliancy; but will appear most vivid in the ray of their natural colour. The green leaves, on the contrary, are of a dark colour, bearing a stronger resemblance to black, than to white; they have, therefore, a greater tendency to absorb, than to reflect rays ; and reflecting very few of any but the blue and yellow rays, they will appear dingy in passing through the other colours of the spectrum.

CAROLINE.

They must, however, reflect great quantities of the green rays, to produce so deep a colour.

MRS. B.

Deepness or darkness of colour proceeds rather from a deficiency than an abundance of reflected rays. Remember that bodies are, of themselves, black; and if a body reflects only a few green rays, it will appear of a dark green; it is the brightness and intensity of the colour which show that a great quantity of rays are reflected.

EMILY.

A white body, then, which reflects all the rays, will appear equally bright in all the colours of the spectrum.

MRS. B.

Certainly.    And this is easily proved by passing a sheet of white paper through the rays of the spectrum.

CAROLINE.

What is the reason that blue often appears green by candle-light?

MRS. B.

The light of a candle is not so pure as that of
s 4

the sun: it has a yellowish tinge, and when refracted by the prism, the yellow rays predominate;
and as blue bodies reflect the yellow rays in the
next proportion (being next in order of refrangibility), the superabundance of yellow rays gives to
blue bodies a greenish hue.

### CAROLINE.

Candle-light must then give to all bodies a yellowish tinge, from the excess of yellow rays; and
yet it is a common remark, that people of a sallow
complexion appear fairer or whiter by candle-light.

### MRS. B.

The yellow cast of their complexion is not so
striking, when every object has a yellow tinge.

### EMILY.

Pray, why does the sun appear red through a
fog?

### MRS. B.

It is supposed to be owing to the red rays having a greater momentum, which gives them power
to traverse so dense an atmosphere. For the same
reason, the sun generally appears red at rising and
setting; as the increased quantity of atmosphere,
which the oblique rays must traverse, loaded with
the mists and vapours which are usually formed at

those times, prevents the other rays from reaching us.

CAROLINE.

And, pray, why are the skies of a blue colour?

MRS. B.

You should rather say, the atmosphere; for the sky is a very vague term, the meaning of which it would be difficult to define philosophically.

CAROLINE.

But the colour of the atmosphere should be white, since all the rays traverse it in their passage to the earth.

MRS. B.

Do not forget that we see none of the rays which pass from the sun to the earth, excepting those which meet our eyes; and this happens only if we look at the sun, and thus intercept the rays, in which case, you know, the sun appears white. The atmosphere is a transparent medium, through which the sun's rays pass freely to the earth; but when reflected back into the atmosphere, their momentum is considerably diminished; and they have not all of them power to traverse it a second time. The momentum of the blue rays is least; these, therefore, are the most

s 5

impeded in their return, and are chiefly reflected by the atmosphere: this reflection is performed in every possible direction; so that whenever we look at the atmosphere, some of these rays fall upon our eyes; hence we see the air of a blue colour. If the atmosphere did not reflect any rays, though the objects on the surface of the earth would be illumined, the skies would appear perfectly black.

CAROLINE.

Oh, how melancholy that would be; and how pernicious to the sight, to be constantly viewing bright objects against a black sky. But what is the reason that bodies often change their colour; as leaves which wither in autumn, or a spot of ink which produces an iron-mould on linen?

MRS. B.

It arises from some chemical change, which takes place in the internal arrangement of the parts, by which they lose their tendency to reflect certain colours, and acquire the power of reflecting others. A withered leaf thus no longer reflects the blue rays; it appears, therefore, yellow, or has a slight tendency to reflect several rays which produce a dingy brown colour.

An ink-spot on linen at first absorbs all the rays; but, exposed to the air, it undergoes a chemical change, and the spot partially regains its tendency

16

to reflect colours, but with a preference to reflect the yellow rays, and such is the colour of the iron-mould.

### EMILY.

Bodies, then, far from being of the colour which they appear to possess, are of that colour which they have the greatest aversion to, which they will not incorporate with, but reject and drive from them.

### MRS. B.

It certainly is so; though I scarcely dare venture to advance such an opinion, whilst Caroline is contemplating her beautiful rose.

### CAROLINE.

My poor rose! you are not satisfied with depriving it of colour, but even make it have an aversion to it; and I am unable to contradict you.

### EMILY.

Since dark bodies absorb more solar rays than light ones, the former should sooner be heated if exposed to the sun?

### MRS. B.

And they are found by experience to be so. Have

s 6

you never observed a black dress to be warmer than a white one?

### EMILY.

Yes, and a white one more dazzling: the black is heated by absorbing the rays, the white dazzling by reflecting them.

### CAROLINE.

And this was the reason that the brown paper was burnt in the focus of the lens, whilst the white paper exhibited the most luminous spot, but did not take fire.

### MRS. B.

It was so.   It is now full time to conclude our lesson.   At our next meeting, I shall give you a description of the eye.

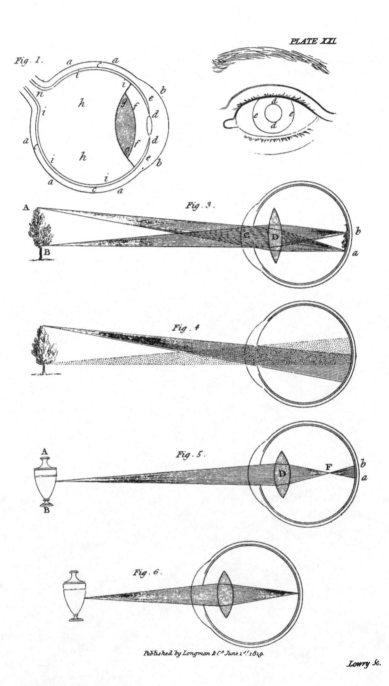

PLATE XXI

Fig. 1.

Fig. 3.

Fig. 4.

Fig. 5.

Fig. 6.

Published by Longman & Cº June 1.ˢᵗ 1819.

Lowry Sc.

# CONVERSATION XVII.

## OPTICS.

## ON THE STRUCTURE OF THE EYE, AND OPTICAL INSTRUMENTS.

DESCRIPTION OF THE EYE. — OF THE IMAGE ON
THE RETINA. — REFRACTION OF THE HUMOURS
OF THE EYE. — OF THE USE OF SPECTACLES.
— OF THE SINGLE MICROSCOPE. — OF THE
DOUBLE MICROSCOPE. — OF THE SOLAR MICRO-
SCOPE. — MAGIC LANTHORN. — REFRACTING
TELESCOPE. — REFLECTING TELESCOPE.

MRS. B.

THE body of the eye is of a spherical form: (fig. I.
Plate XXI.) it has two membranous coverings; the
external one, *a a a*, is called the sclerotica: this
has a projection in that part of the eye which is

exposed to view, *b b*, which is called the cornea, because, when dried, it has nearly the consistence of very fine horn, and is sufficiently transparent for the light to obtain free passage through it.

The second membrane which lines the cornea, and envelopes the eye, is called the choroid, *c c c*; this has an opening in front, just beneath the cornea, which forms the pupil, *d d*, through which the rays of light pass into the eye. The pupil is surrounded by a coloured border, called the iris, *e e*, which, by its muscular motion, always preserves the pupil of a circular form, whether it is expanded in the dark, or contracted by a strong light. This you will understand better by examining fig. 2.

### EMILY.

I did not know that the pupil was susceptible of varying its dimensions.

### MRS. B.

The construction of the eye is so admirable, that it is capable of adapting itself, more or less, to the circumstances in which it is placed. In a faint light the pupil dilates so as to receive an additional quantity of rays, and in a strong light it contracts, in order to prevent the intensity of the light from injuring the optic nerve. Observe Emily's eyes, as she sits looking towards the windows: her pupils appear very small, and the iris large. Now, Emily, turn from the light, and cover your eyes with your hand, so as entirely to exclude it for a few moments.

### CAROLINE.

How very much the pupils of her eyes are now enlarged, and the iris diminished. This is, no doubt, the reason why the eyes suffer pain, when from darkness they suddenly come into a strong light; for the pupil being dilated, a quantity of rays must rush in before it has time to contract.

### EMILY.

And when we go from a strong light into obscurity, we at first imagine ourselves in total darkness; for a sufficient number of rays cannot gain admittance into the contracted pupil, to enable us to distinguish objects: but in a few minutes it dilates, and we clearly perceive objects which were before invisible.

### MRS. B.

It is just so. The choroid *c c*, is imbued with a black liquor, which serves to absorb all the rays that are irregularly reflected, and to convert the body of the eye into a more perfect camera obscura. When the pupil is expanded to its utmost extent, it is capable of admitting ten times the quantity of light that it does when most contracted. In cats, and animals which are said to see in the dark, the

power of dilatation and contraction of the pupil is
still greater : it is computed that their pupils may
receive one hundred times more light at one time
than at another.

Within these coverings of the eye-ball are contained three transparent substances, called humours.
The first occupies the space immediately behind the
cornea, and is called the aqueous humour, $f f$,
from its liquidity and its resemblance to water.
Beyond this is situated the crystalline humour, $g g$,
so called from its clearness and transparency : it has
the form of a lens, and refracts the rays of light in
a greater degree of perfection than any that have
been constructed by art : it is attached by two muscles, $m m$, to each side of the choroid.  The back
part of the eye, between the crystalline humour and
the retina, is filled by the vitreous humour, $h h$,
which derives its name from a resemblance it is supposed to bear to glass or vitrified substances.

The membranous coverings of the eye are intended chiefly for the preservation of the retina, $i i$,
which is by far the most important part of the eye,
as it is that which receives the impression of the
objects of sight, and conveys it to the mind.  The
retina consists of an expansion of the optic nerve,
of a most perfect whiteness : it proceeds from the
brain, enters the eye, at $n$, on the side next the
nose, and is finely spread over the interior surface
of the choroid.

The rays of light which enter the eye by the pupil are refracted by the several humours in their passage through them, and unite in a focus on the retina.

CAROLINE.

I do not understand the use of these refracting humours : the image of objects is represented in the camera obscura, without any such assistance.

MRS. B.

That is true; but the representation would be much more strong and distinct, if we enlarged the opening of the camera obscura, and received the rays into it through a lens.

I have told you that rays proceed from bodies in all possible directions. We must, therefore, consider every part of an object which sends rays to our eyes, as points from which the rays diverge, as from a centre.

EMILY.

These divergent rays, issuing from a single point, I believe you told us, were called a pencil of rays?

MRS. B.

Yes. Now, divergent rays, on entering the pupil, do not cross each other; the pupil, however, is sufficiently large to admit a small pencil of them; and these, if not refracted to a focus by the humours, would continue diverging after they had passed the pupil, would fall dispersed upon the retina, and thus the image of a single point would be expanded over a large portion of the retina. The divergent rays from every other point of the object would be spread over a similar extent of space, and would interfere and be confounded with the first; so that no distinct image could be formed, and the retina would represent total confusion both of figure and colour. Fig. 3. represents two pencils of rays issuing from two points of the tree A B, and entering the pupil C, refracted by the crystalline humour D, and forming distinct images of the spot they proceed from, on the retina, at *a b*. Fig. 4. differs from the preceding, merely from not being supplied with a lens; in consequence of which the pencils of rays are not refracted to a focus, and no distinct image is formed on the retina. I have delineated only the rays issuing from two points of an object, and distinguished the two pencils in fig. 4. by describing one of them with dotted lines: the interference of these two pencils of rays on the retina will enable you to form an idea of the confusion which would arise, from thousands and millions of points at the same instant pouring their divergent rays upon the retina.

EMILY.

True; but I do not yet well understand how the refracting humours remedy this imperfection.

### MRS. B.

The refraction of these several humours unite the whole of a pencil of rays, proceeding from any one point of an object, to a corresponding point on the retina, and the image is thus rendered distinct and strong. If you conceive, in fig. 3., every point of the tree to send forth a pencil of rays similar to those, A B, every part of the tree will be as accurately represented on the retina as the points *a b.*

### EMILY.

How admirably, how wonderfully, this is contrived!

### CAROLINE.

But since the eye requires refracting humours in order to have a distinct representation formed on the retina, why is not the same refraction necessary for the image formed in the camera obscura?

### MRS. B.

Because the aperture through which we received the rays into the camera obscura is extremely small; so that but very few of the rays diverging from a point gain admittance; but we will now enlarge the aperture, and furnish it with a lens, and you will find the landscape to be more perfectly represented.

### CAROLINE.

How obscure and confused the image is now that you have enlarged the opening, without putting in the lens.

### MRS. B.

Such, or very similar, would be the representation on the retina, unassisted by the refracting humours. But see what a difference is produced by the introduction of the lens, which collects each pencil of divergent rays into their several foci.

### CAROLINE.

The alteration is wonderful : the representation is more clear, vivid, and beautiful than ever.

### MRS. B.

You will now be able to understand the nature of that imperfection of sight, which arises from the eyes being too prominent. In such cases, the crystalline humour, D, (fig. 5.) being extremely convex, refracts the rays too much, and collects a pencil, proceeding from the object A B, into a focus, F, before they reach the retina. From this focus, the rays proceed diverging, and consequently form a very confused image on the retina, at $a\,b$. This is the defect of short-sighted people.

### EMILY.

I understand it perfectly. But why is this defect

PLATE XXII.

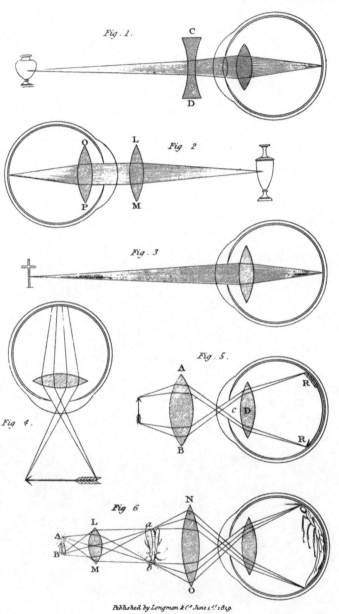

*Fig. 1.*

C
D

*Fig 2*

O
L
P
M

*Fig. 3*

*Fig. 4.*

*Fig. 5.*

A
B
c D
R
R

*Fig 6*

N
L
A
B
M
O

Published by Longman & Co. June 1st 1819.

Lowry Sc.

remedied by bringing the object nearer to the eye, as we find to be the case with short-sighted people?

MRS. B.

The nearer you bring an object to your eye, the more divergent the rays fall upon the crystalline humour, and they are consequently not so soon converged to a focus: this focus, therefore, either falls upon the retina, or at least approaches nearer to it, and the object is proportionably distinct, as in fig. 6.

EMILY.

The nearer, then, you bring an object to a lens, the further the image recedes behind it.

MRS. B.

Certainly. But short-sighted persons have another resource for objects which they cannot approach to their eyes; this is to place a concave lens, C D, (fig. 1. Plate XXII.) before the eye, in order to increase the divergence of the rays. The effect of a concave lens is, you know, exactly the reverse of a convex one: it renders parallel rays divergent, and those which are already divergent, still more so. By the assistance of such glasses, therefore, the rays from a distant object fall on the pupil, as divergent as those from a less distant object; and, with short-sighted people, they throw the image of a distant object back as far as the retina.

CAROLINE.

This is an excellent contrivance, indeed.

MRS. B.

And tell me, what remedy would you devise for such persons as have a contrary defect in their sight; that is to say, in whom the crystalline humour, being too flat, does not refract the rays sufficiently, so that they reach the retina before they are converged to a point?

CAROLINE.

I suppose that a contrary remedy must be applied to this defect; that is to say, a convex lens, L M, fig. 2., to make up for the deficiency of convexity of the crystalline humour, O P. For the convex lens would bring the rays nearer together, so that they would fall either less divergent, or parallel on the crystalline humour; and, by being sooner converged to a focus, would fall on the retina.

MRS. B.

Very well, Caroline. This is the reason why elderly people, the humours of whose eyes are decayed by age, are under the necessity of using convex spectacles. And when deprived of that resource, they hold the object at a distance from their eyes, as in fig. 4, in order to bring the focus forwarder.

### CAROLINE.

I have often been surprised, when my grandfather reads without his spectacles, to see him hold the book at a considerable distance from his eyes. But I now understand it; for the more distant the object is from the crystalline, the nearer the image will be to it.

### EMILY.

I comprehend the nature of these two opposite defects very well; but I cannot now conceive, how any sight can be perfect: for if the crystalline humour is of a proper degree of convexity, to bring the image of distant objects to a focus on the retina, it will not represent near objects distinctly; and if, on the contrary, it is adapted to give a clear image of near objects, it will produce a very imperfect one of distant objects.

### MRS. B.

Your observation is very good, Emily; and it is true, that every person would be subject to one of these two defects, if we had it not in our power to increase or diminish the convexity of the crystalline humour, and to project it towards, or draw it back from the object, as circumstances require. In a young well-constructed eye, the two muscles to which the crystalline humour is attached have 'so perfect a command over it, that the focus of the rays constantly falls on the retina, and an image is formed equally distinct both of distant objects and of those which are near.

CAROLINE.

In the eyes of fishes, which are the only eyes I have ever seen separate from the head, the cornea does not protrude, in that part of the eye which is exposed to view.

MRS. B.

The cornea of the eye of a fish is not more convex than the rest of the ball of the eye; but to supply this deficiency, their crystalline humour is spherical, and refracts the rays so much, that it does not require the assistance of the cornea to bring them to a focus on the retina.

EMILY.

Pray, what is the reason that we cannot see an object distinctly, if we approach it very near to the eye?

MRS. B.

Because the rays fall on the crystalline humour too divergent to be refracted to a focus on the re-

tina, the confusion, therefore, arising from viewing
an object too near the eye, is similar to that which
proceeds from a flattened crystalline humour; the
rays reach the retina before they are collected to a
focus, (fig. 4.) If it were not for this imperfection,
we should be able to see and distinguish the parts
of objects, which are now invisible to us from their
minuteness; for could we approach them very near
the eye, their image on the retina would be so much
magnified as to render them visible.

EMILY.

And could there be no contrivance to convey the
rays of objects viewed close to the eye, so that they
should be refracted to a focus on the retina?

MRS. B.

The microscope is constructed for this purpose.
The single microscope (fig. 5.) consists simply of
a convex lens, commonly called a magnifying glass;
in the focus of which the object is placed, and
through which it is viewed: by this means, you are
enabled to approach your eye very near the object,
for the lens A B, by diminishing the divergence of
the rays, before they enter the pupil C, makes them
fall parallel on the crystalline humour D, by which
they are refracted to a focus on the retina, at R R.

T

### EMILY.

This is a most admirable invention, and nothing can be more simple, for the lens magnifies the object merely by allowing us to bring it nearer to the eye.

### MRS. B.

Those lenses, therefore, which have the shortest focus will magnify the object most, because they enable us to bring the object nearest to the eye.

### EMILY.

But a lens, that has the shortest focus, is most bulging or convex; and the protuberance of the lens will prevent the eye from approaching very near to the object.

### MRS. B.

This is remedied by making the lens extremely small: it may then be spherical without occupying much space, and thus unite the advantages of a short focus, and of allowing the eye to approach the object.

### CAROLINE.

We have a microscope at home, which is a much more complicated instrument than that you have described.

PLATE XXIII.

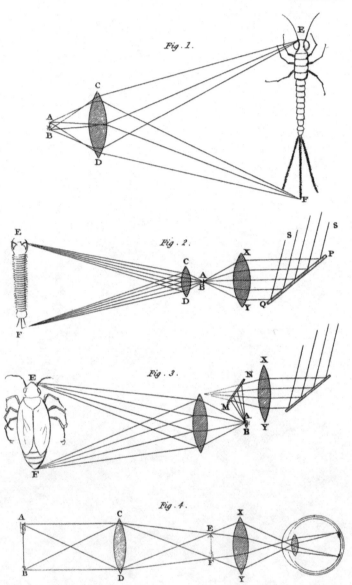

Fig. 1.

Fig. 2.

Fig. 3.

Fig. 4.

Published by Longman & Co. June 1st 1819.

Lowry Sc.

It is a double microscope (fig. 6.), in which you see, not the object A B, but a magnified image of it, *a b.* In this microscope, two lenses are employed, the one, L M, for the purpose of magnifying the object, is called the object glass; the other, N O, acts on the principle of the single microscope, and is called the eye-glass.

There is another kind of microscope, called the solar microscope, which is the most wonderful from its great magnifying power: in this we also view, an image formed by a lens, not the object itself. As the sun shines, I can show you the effect of this microscope; but for this purpose, we must close the shutters, and admit only a small portion of light, through the hole in the window-shutter, which we used for the camera obscura. We shall now place the object A B, (Plate XXIII. fig. 1.) which is a small insect, before the lens C D, and nearly at its focus: the image E F, will then be represented on the opposite wall in the same manner as the landscape was in the camera obscura; with this difference, that it will be magnified, instead of being diminished. I shall leave you to account for this, by examining the figure.

MRS. B.

EMILY.

I see it at once. The image E F is magnified, because it is farther from the lens, than the object

A B; while the representation of the landscape was diminished, because it was nearer the lens, than the landscape was. A lens, then, answers the purpose equally well, either for magnifying or diminishing objects?

MRS. B.

Yes: if you wish to magnify the image, you place the object near the focus of the lens; if you wish to produce a diminished image, you place the object at a distance from the lens, in order that the image may be formed in, or near the focus.

CAROLINE.

The magnifying power of this microscope, is prodigious; but the indistinctness of the image for want of light, is a great imperfection. Would it not be clearer, if the opening in the shutter were enlarged, so as to admit more light.

MRS. B.

If the whole of the light admitted does not fall upon the object, the effect will only be to make the room lighter, and the image consequently less distinct.

EMILY.

But could you not by means of another lens bring a large pencil of rays to a focus on the object, and thus concentrate the whole of the light admitted upon it?

MRS. B.

Very well. We shall enlarge the opening, and place the lens X Y (fig. 2.) in it, to converge the rays to a focus on the object A B. There is but one thing more wanting to complete the solar microscope, which I shall leave to Caroline's sagacity to discover.

CAROLINE.

Our microscope has a small mirror attached to it, upon a moveable joint, which can be so adjusted as to receive the sun's rays, and reflect them upon the object: if a similar mirror were placed to reflect light upon the lens, would it not be a means of illuminating the object more perfectly.

MRS. B.

You are quite right. P Q (fig. 2.) is a small mirror, placed on the outside of the window-shutter, which receives the incident rays S S, and reflects them on the lens X Y. Now that we have completed the apparatus let us examine the mites on this piece of cheese, which I place near the focus of the lens.

T 3

CAROLINE.

Oh, how much more distinct the image now is, and how wonderfully magnified! The mites on the cheese look like a drove of pigs scrambling over rocks.

EMILY.

I never saw any thing so curious. Now, an immense piece of cheese has fallen: one would imagine it an earthquake: some of the poor mites must have been crushed; how fast they run, — they absolutely seem to gallop.

But this microscope can be used only for transparent objects; as the light must pass through them to form the image on the wall?

MRS. B.

Very minute objects, such as are viewed in a microscope, are generally transparent; but when opaque objects are to be exhibited, a mirror M N (fig. 3.) is used to reflect the light on the side of the object next the wall: the image is then formed by light reflected from the object, instead of being transmitted through it.

EMILY.

Pray, is not a magic lanthorn constructed on the same principles?

Yes; with this difference, that the light is supplied by a lamp, instead of the sun.

The microscope is an excellent invention, to enable us to see and distinguish objects, which are too small to be visible to the naked eye. But there are objects, which, though not really small, appear so to us, from their distance; to these we cannot apply the same remedy; for when a house is so far distant, as to be seen under the same angle as a mite which is close to us, the effect produced on the retina is the same: the angle it subtends is not large enough for it to form a distinct image on the retina.

EMILY.

Since it is impossible, in this case, to approach the object to the eye, cannot we by means of a lens bring an image of it nearer to us?

MRS. B.

Yes; but then, the object being very distant from the focus of the lens, the image would be too small to be visible to the naked eye.

EMILY.

Then, why not look at the image through another lens, which will act as a microscope, enable us to bring the image close to the eye, and thus render it visible?

Very well, Emily; I congratulate you on having invented a telescope. In figure 4, the lens C D, forms an image E F, of the object A B; and the lens X Y serves the purpose of magnifying that image; and this is all that is required in a common refracting telescope.

But in fig. 4. the image is not inverted on the retina, as objects usually are: it should therefore appear to us inverted; and that is not the case in the telescopes I have looked through.

When it is necessary to represent the image erect, two other lenses are required; by which means a second image is formed, the reverse of the first, and consequently upright. These additional glasses are used to view terrestrial objects; for no inconvenience arises from seeing the celestial bodies inverted.

The difference between a microscope and a telescope, seems to be this : — a microscope produces a magnified image, because the object is nearest the lens; and a telescope produces a diminished image, because the object is furthest from the lens.

Your observation applies only to the lens C D, or object-glass, which serves to bring an image of the object nearer the eye; for the lens X Y, or eye-glass, is, in fact, a microscope, as its purpose is to magnify the image.

When a very great magnifying power is required, telescopes are constructed with concave mirrors, instead of lenses. Concave mirrors, you know, produce by reflection, an effect similar to that of convex lenses by refraction. In reflecting telescopes, therefore, mirrors are used in order to bring the image nearer the eye; and a lens or eye-glass the same as in the refracting telescope to magnify the image.

The advantage of the reflecting telescope is, that mirrors whose focus is six feet will magnify as much as lenses of a hundred feet.

CAROLINE.

But I thought it was the eye-glass only which magnified the image; and that the other lens served to bring a diminished image nearer to the eye.

The image is diminished in comparison to the object, it is true; but it is magnified if you compare it to the dimensions of which it would appear without the intervention of any optical instrument; and this magnifying power is greater in reflecting than in refracting telescopes.

We must now bring our observations to a conclusion, for I have communicated to you the whole of my very limited stock of knowledge of Natural Philosophy. If it will enable you to make further progress in that science, my wishes will be satisfied; but remember that, in order that the study of nature may be productive of happiness, it must lead to an entire confidence in the wisdom and goodness of its bounteous Author.

# INDEX.

U

424                    INDEX.

## T

THE END.

Printed by Strahan and Spottiswoode,
Printers-Street, London.

Printed in the United States
By Bookmasters